German Seaplanes of WWI

Sablatnig, Kaiserliche Werften, Lübeck-Travemünde, LTG, & Oertz

A Centennial Perspective on Great War Airplanes

Jack Herris

Great War Aviation Centennial Series #15

This book is dedicated to Martin Digmayer in appreciation for all his excellent scale drawings that greatly enhance this book series.

Acknowledgements

My sincere thanks to Colin Owers, Aaron Weaver, Bill Toohey, Greg VanWyngarden, Henning Oppermann, and Reinhard Zankl for photographs and helpful suggestions and material. Thanks to Dave Hooper for the good quality SVK drawings and to Bob Pearson for his color profiles, Martin Digmayer for scale drawings, and the Deutsches Technikmuseum in Berlin and the Museum of Flight in Seattle for photographs, and the kind assistance of Cross & Cockade International. This Update 1 is thanks to Henning Oppermann, who kindly provided the photos of the Oertz W7. Any errors are my responsibility.

The cover painting **The Seahawks** by Steve Anderson illustrates an attack by Sablatnig SF5 Marine #1023 on a Russian steamer on August 29, 1917. On that day *Oblt. dRMI* Hermann Pohrt and *Flgmr.d.R.* Johannes Jensen, flying from Angersee, spotted two loaded Russian cargo steamers. They attacked the first steamer, about 1,000 tons, with six bombs from about 700 meters altitude. Three hits were seen that caused an explosion and fire amidships. The steamer immediately turned east and lowered a lifeboat with 12–14 crewmen. Several minutes later large portions of the mid and after section of the ship were seen to be burning. Steve's dramatic painting of this event was the cover for *Over the Front*, Winter 2013. Please see Steve's website at: www.anderson-art.com

Color aircraft profiles © Bob Pearson. Purchase his CD of WWI aircraft profiles for $50 US/Canadian, 40 €, or £30, airmail postage included, via Paypal to Bob at: bpearson@kaien.net

For our aviation books in print and electronic format, please see our website at: www.aeronautbooks.com. I am looking for photographs of the less well-known German aircraft of WWI. For questions or to help with photographs you may contact me at jherris@me.com.

Interested in WWI aviation? Join The League of WWI Aviation Historians (www.overthefront.com) and Cross & Cockade International (www.crossandcockade.com).

ISBN: 978-1-935881-27-8
© 2015 Aeronaut Books, all rights reserved
Text © 2015 Jack Herris
Design and layout: Jack Herris
Cover design: Jack Herris
Digital photo editing: Jack Herris

www.aeronautbooks.com

Table of Contents

Sablatnig Floatplanes	3	**LTG Floatplanes**	70
Sablatnig SF1	6	LTG FD 1	70
Sablatnig SF2	8	**Oertz Flying Boats**	74
Sablatnig SF3	19	Oertz F.B.1	75
Sablatnig SF4	21	Oertz F.B.2	76
Sablatnig SF5	25	Oertz F.B.3	77
Sablatnig SF6 / B.I	34	Oertz W4	79
Sablatnig SF7	36	Oertz W5	81
Sablatnig SF8	37	Oertz W6 *Flugschoner*	85
Kaiserliche Werften Floatplanes	42	Oertz W7	90
KW Type 401	42	Oertz W8	91
KW Type 462	47	**In Retrospect**	94
KW Type 467	51	**Bibliography**	95
KW Type 945	51	**Index**	95
KW Type 947	51	**Scale Drawings (MD)**	
KW Type 1105	52	Sablatnig SF2 (1/48)	96
KW Type 1650	55	Sablatnig SF5 (1/48)	100
Lübeck-Travemünde Floatplanes	56	Lübeck-Travemünde F2 #766 (1/72)	103
Lübeck-Travemünde F1	56	Lübeck-Travemünde F2 #1150 (1/72)	106
Lübeck-Travemünde F2	57	Lübeck-Travemünde F4 (1/72)	109
Lübeck-Travemünde F3	64	Sablatnig C.III (1/72)	112
Lübeck-Travemünde F4	65	**Afterword**	114

Forward

The purpose of this book is to document the little-known German floatplanes built by Sablatnig, the Kaiserliche Werften, Lübeck-Travemünde, and LTG plus the Oertz flying boats. The few Sablatnig land airplanes are covered in *Nachtflugzeug! German N-Types of WWI*. The Sablatnig C.III drawing was not available when that book was written so is included here for completeness.

Below: Lübeck-Travemünde F2 Marine Number 1978 was the next to last F2 ordered and is the last for which delivery is confirmed in SVK records, which are incomplete after June 1918.

Sablatnig Floatplanes

Above: Joseph Sablatnig (rear seat) giving flight instruction, possibly at the Wright Flying Machine, Ltd., school.

Joseph Sablatnig was born February 9, 1886 in Klagenfurt, Graz, then part of the Austro-Hungarian Empire. After graduating from high school in 1904 in Klagenfurt, he studied mechanical engineering in Brno, and graduated as an engineer in December 1909. In late 1909 he took pilot training with the Wright Flying Machine, Ltd., near Berlin and in October 1910 received Austrian pilot's license No. 12. He participated in Austria's first flying event it May 1910. He also achieved his Ph.D. in Vienna in 1910. In 1911 he flew one of the first flights at night. In 1912 he completed his first airplane in Vienna, and later that year became a German citizen. He went on to win a number of flying awards and set a world altitude record in 1913 at Johannisthal airfield in Berlin.

In 1913 Sablatnig became a director of Union Flugzeugwerke GmbH, where he did technical work and flying. When the Union company went bankrupt in 1915, he decided to start his own airplane design and manufacturing company.

In October 1915 he founded Sablatnig Flugzeugbau GmbH in Berlin, which had about a thousand employees by the end of the war. Most of his designs were single-engine biplane floatplanes, although his company also produced some prototype C-types and the production Sablatnig N.I night bomber, all of which were single-engine aircraft.

All Sablatnig seaplanes were fairly conventional designs that utilized the wire-braced, fabric-covered wood construction typical for the period. Sablatnig continued to use these materials for his land-based

SVK Table of Sablatnig Seaplane Orders and Deliveries

Order Number	Type	Marine Numbers	Design	Class & Engine	1914 J F M A M J J A S O N D	1915 J F M A M J J A S O N D
1	490	490	SF1	B 160M		① 1
2	609	580/585	SF2	BFT 160M		⑥
3	609	608/618	SF2	BFT 160M		⑩
4		619	SF3	CFT 220B		
5	609	705/714	SF2	BFL 160M	L.F.G.	
6		968/987	SF5	HFT 150B		
7		1017/1036	SF5	B 150B	L.V.G.	
8	1230	1224/1233	SF5	HFT 150B		
9	1361	1352/1371	SF5	HFT 150B		
10	1476	1475/1477	SF7	C2MG 240Mb		
11	1361	1514	SF5	BFT 150B		
12	2021	2020/2022	SF8	School 150B		
13	FF1822	1842/1856	FF59c	CHFT 200B	L.F.G. FF License	
14	FF1822	1872/1901	SF8	CHFT 200B	FF License	
15	2021	6001/6030	SF8	School 150B		
16		6031/6050	FF49c	CHFT 200B		
				Orders that Month		1 6 10
				Deliveries that Month		1
				Orders that Year		17
Orders: ③	Deliveries: 2			Deliveries that Year		1

designs, although the last two types, the C.II and C.III, featured plywood-covered fuselages.

After the war he built the Sablatnig P.1 four-seat civil aircraft that made the first international flight to Stockholm, Sweden, and started a small airline in March 1919. In October 1920 he merged his airline with another small airline. His Sablatnig P.3 civil aircraft was used by Lufthansa into the early 1930s. He also designed and built cars.

Sablatnig was captured by Soviet forces in Berlin on June 16, 1945 and imprisoned in the former Buchenwald concentration camp, where he died February 28, 1946.

Sablatnig SF-Series Specifications

	SF1	SF2 (#609–618)	SF4 (#900)
Engine	160 hp Mercedes D.III	160 hp Mercedes D.III	150 hp Benz Bz.III
Span	19.1 m	18.53 m	12.0 m
Length	—	9.525 m	8.33 m
Wing Area	—	56 m²	28.26 m²
Wt. Empty	1,105 kg	1,078 kg	798 kg
Wt. Loaded	1,650 kg	1,697 kg	1,078 kg
Speed	125 km/h	130 km/h	158 km/h
Climb: 1,000 m	10 minutes	—	5.5 minutes
1,500 m	—	18 minutes	—
2,000 m	—	—	14 minutes
Armament	None	None	1 Spandau
Notes: 1: SF3 was powered by a 220 hp Benz Bz.IV; no other data available. 2: SF4 #901 (triplane) spanned 9.25 m, was 8.33 m long, and had a wing area of 28.38 m².			

Sablatnig SF-Series Specifications

	SF5 (#1361)	SF6	SF7	SF8
Engine	150 hp Benz Bz.III	150 hp Benz Bz.III	240 hp Maybach Mb.IV	150 hp Benz Bz.III
Span	17.3 m	17.3 m	—	16.0 m
Length	9.6 m	8.3 m	—	10.2 m
Wing Area	50.5 m²	—	—	54.6 m²
Wt. Empty	1,052 kg	—	—	1,183 kb
Wt. Loaded	1,605 kg	—	2,120 kg	1,574 kg
Speed	148 km/h	—	162 km/h	130 km/h
Climb: 1,000 m	11.6 minutes	—	8 minutes	14.7 minutes
1,500 m	—	—	—	—
2,000 m	21.8 minutes	—	—	24.8 minutes
3,000 m	—	—	36 min.	—
Armament	None	None	1 Spandau & 1 Parabellum	None

Sablatnig SF1

Above: This side view of the sole SF1 shows the very streamlined nose and rakish lines to advantage. The engine was a 160 hp Mercedes D.III.

The Sablatnig SF1 was designed as an unarmed, two-seat reconnaissance floatplane of conventional design and wire-braced, fabric-covered wood construction typical of the time. Powered by a 160 hp Mercedes D.III engine, only one airframe, Marine #490, was built. Delivered to the SVK (*SVK – Seeflugzeug-Versuchskommando*) on 26 August 1915, tests were satisfactory and it was transferred to Borkum in November. The SF1 flew several operational patrols from Borkum in December. On February 3, 1916, the SF1 failed to return from a patrol and no trace of the floatplane was found. Production orders for a modified SF1, the SF2, followed.

Sablatnig SF1 Marine #490

Above: The front view of the SF1 emphasizes the care taken in minimizing frontal area for minimum drag. Unfortunately, the myriad of bracing wires added significant drag to the airframe, negating much of the advantage of the streamlining.

Below: The rear quarter view of the SF1 emphasizes its streamlined vertical tail surfaces – and the struts bracing it. As a mechanical engineer and pilot Sablatnig understood the need for streamlining but was unable to design a robust structure without excessive drag. In fact, the structural design was inadequate both in strength and aerodynamic cleanliness despite the well-shaped nose, as exemplified in the need for additional bracing struts and wires above the outer interplane struts.

Sablatnig SF2

Above: This side view of the SF2 Marine #580, the first production SF2, shows the streamlined nose and rakish profile of the SF2 and its enlarged fin, the major change from the SF1. The engine was a 160 hp Mercedes D.III.

The SF2 was derived from the earlier SF1 by enlarging the vertical tail surfaces for improved directional stability and modifying the wing bracing by deleting a diagonal strut on each side. Like the SF1, the SF2 was an unarmed, two-seat reconnaissance floatplane; however, a radio transmitter was carried, making it a naval category BFT. Powered by the 160 hp Mercedes D.III as was the SF1, the SF2 was slightly lighter and faster than the SF1.

The SF2 was put into production for the German navy and 36 were produced in the following batches and Marine Numbers:

First production batch: Marine Nr: 580–585 (6 aircraft)
Second production batch: Marine Nr: 609–618 (10 aircraft)
Third production batch: Marine Nr: 705–714 (10 aircraft)
Fourth production batch: Marine Nr: 791–800 (LVG built) (10 aircraft)

The first production SF2 floatplanes were delivered between June and September 1916. The SF2 was used both for operational reconnaissance missions and training.

Above & Below: The first SF2, Marine #580, embodies the streamlined nose and rakish profile typical of early Sablatnig seaplanes, showing that Sablatnig understood the need for streamlining to achieve performance. The SF2 also exhibits the profusion of struts and bracing wires typical of Sablatnig seaplanes, revealing how difficult it was to design a sturdy yet low-drag airframe with the materials and design techniques of the time. Despite Sablatnig's advanced technical training, he was unable to advance the structural and aerodynamic state of the art for seaplanes. The result was the SF series had, at best, average performance and mediocre strength and durability.

Above & Below: More views of SF2 Marine #580, the first production SF2. Despite the struts and many bracing wires, the wing cellule was not overly robust.

Above & Below: More views of SF2 Marine #580, the first production SF2. It carries the *Marine* recognition pennants on the lower wingtips. A Brandenburg GW is in the right background in the lower photo.

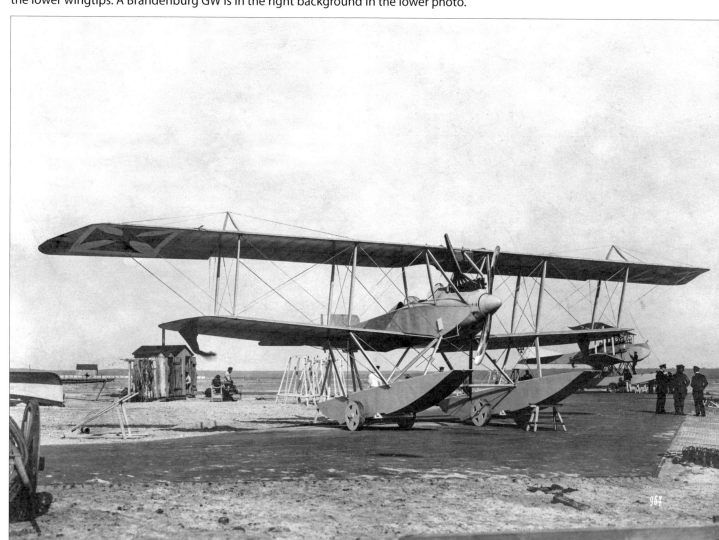

Sablatnig SF2 SVK Drawing

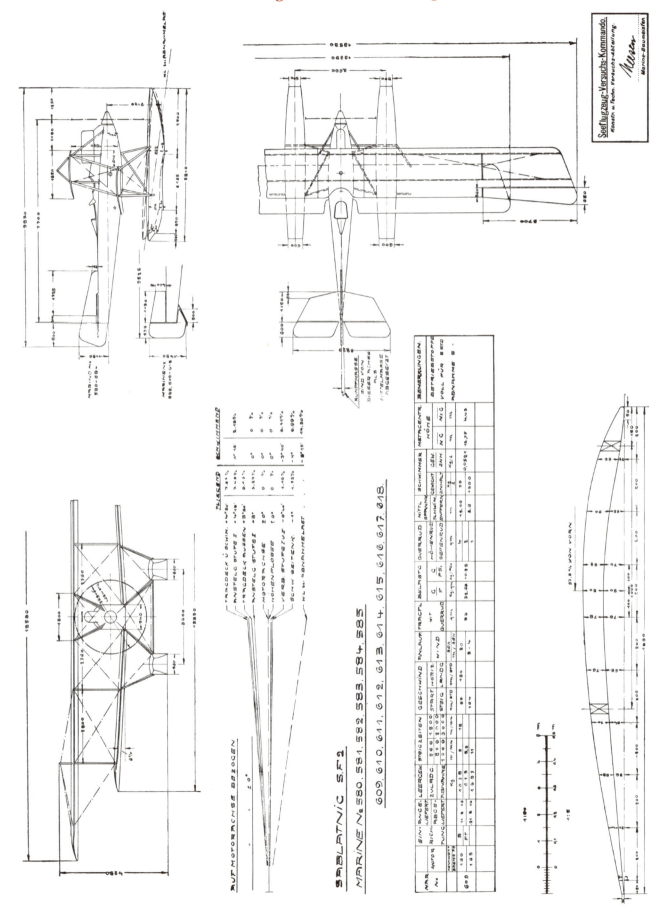

Above: This view of an SF2, likely Marine #580, depicts the attention taken to minimize frontal area to eliminate drag.
Below: SF2 Marine #580, the first production SF2, carries the *Marine* recognition pennants on the lower wingtips.

Above: SF2 Marine #795 was part of the last SF2 production batch ordered. This batch was built by LVG, which enlarged the vertical tail surfaces for improved directional stability.

Above: SF2 Marine #799 was also part of the last SF2 production batch built by LVG. In addition to the enlarged tail featured by LVG-built machines it carried banded camouflage on the fuselage and floats. SF2 floatplanes wore national insignia under the extended tips of the upper wings rather than under the lower wings as commonly done.

Above & Below: SF2 Marine #799 was the next to last SF2 ordered. In addition to its banded camouflage on the fuselage and floats it also carries the *Marine* identification pennants attached to the lower wingtips.

Above: SF2 Marine #612 (at left) and another SF2 at *SFS Angersee* warm their engines before departing on a patrol over the Baltic Sea.

Below: SF2 Marine #612 at *SFS Angersee* is shown after suffering damage to the floats, struts, and airframe when the starboard landing gear collapsed during landing on September 28, 1916. The floats and struts had been weakened during the takeoff run following a rescue attempt of one of the *Torpedostaffel's* crew members from damaged Brandenburg GW *T4*, a twin-engine torpedo floatplane.

Above: *Kaiserin* Augusta-Viktoria inspects an SF2 during her visit to the naval air station at Kiel-Holtenau with Prinz Heinrich. One of the guests was *Kapitänleutnant* Wolfgang Plüschow, the famous *Flieger of Tsingtao*.

Below: Marine #580 at the *SVK* facility at Warnemünde appears to be the heavily-photographed SF2,

Above: Taken at Stralsund in May 1917, this photo shows a group of men attending a photography course. The bracket on the side of the SF2 fuselage is for the wind-driven generator for the wireless transmitter.

Below: This SF2 is running up its engine prior to flight.

Sablatnig SF3

Above & Below: The Sablatnig SF3 was compact for a two-seat fighter. However, the SF3 featured a profusion of struts and bracing wires that certainly created more drag than the cleverly-designed W12 and only one prototype was built.

Sablatnig produced two two-seat floatplane fighter designs. First was the SF3 shown here, a sturdy-looking aircraft powered by a 200 hp Benz Bz.IV. Armament was one fixed, forward-firing gun for the pilot and a flexible gun for the observer.

The relatively low drag of the streamlined, ply-covered fuselage was more than compensated for by its multitude of struts and bracing wires, flying qualities were unsatisfactory, and it remained a single prototype. No further details are known.

Above: The Sablatnig SF3 prototype was finished in the standard late-war naval camouflage.

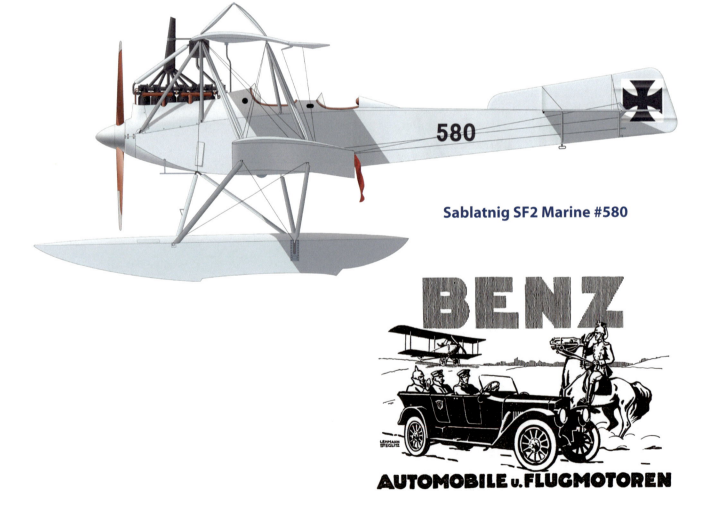

Sablatnig SF2 Marine #580

Sablatnig SF4 Biplane & Triplane

Above & Below: The sole Sablatnig SF4 biplane prototype was Marine Number 900.

Designed as a single-seat floatplane fighter, the SF4 was unique in that it was built in both biplane and triplane versions. Both were powered by a 150 hp Benz Bz.III and carried one fixed, forward-firing Spandau machine gun. The biplane, Marine #900, was tested first, and while speed was competitive, it had the lowest climb rate of all the single-seat floatplane fighter competitors. Worse, its maneuverability was poor due to its large wingspan and, despite its multitude of bracing wires, its structure was insufficiently robust; wing vibration was excessive in even a shallow dive. To improve the climb rate a triplane version, Marine #901, was built; like the biplane it was not competitive.

Company founder Josef Sablatnig was a trained mechanical engineer and a well-known pioneer pilot, so it is especially disappointing that he was unable to design an airframe that was at once light, strong, and streamlined. Although nose entry was streamlined, the wing structure created a lot of drag due to extensive bracing wires, and despite that the wing structure was weak.

Above: The SF4 biplane prototype photographed in the snow; the broad interplane struts may have helped streamlining but obstructed the pilot's field of view to the sides.

Below: The SF4 triplane carried Marine #901 and had ailerons on all wings. Little information is available on the triplane SF4, but it was not selected for production.

Above: Apparently the performance of the SF4 triplane was no improvement over the SF4 biplane and, like the biplane, only one was built.

Sablatnig SF2 Marine #799

Sablatnig SF4 SVK Drawing

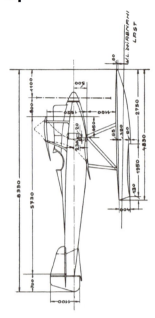
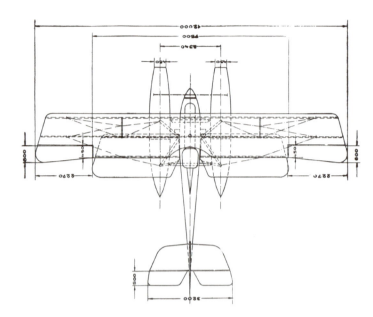

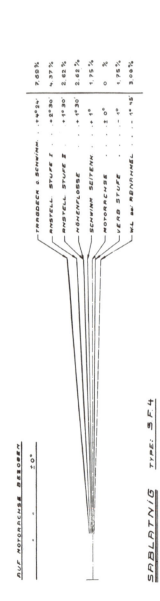
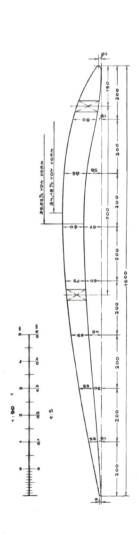

Sablatnig SF5

Above: Tiger-striped Marine Number 968 was the first SF5. The SF5 was closely based on the earlier SF2, although minor structural modifications were made to reduce weight. The most obvious change was the revised wing struts; the overhead bracing of the upper wing was replaced by an additional pair of struts attached to the front, outboard interplane strut. The 160 hp Mercedes used in the SF2 was in great demand for fighters, so the SF5 used the 150 hp Benz Bz.III.

Based on the modest success of the earlier SF2, the SF5 was developed to replace it in production. Like the SF2, the SF5 was an unarmed two-seat reconnaissance floatplane with wireless transmitter, naval category BFT. The SF5 was powered by the 150 hp Benz Bz.III instead of the 160 hp Mercedes D.III used in the SF2, probably because fighter aircraft had priority for the preferred Mercedes. The SF5 also had revised wing bracing; an additional pair of struts bracing the outer wingtips of the upper wing replaced the pylon struts over the outboard bay of interplane struts. This gave the SF5 a sleeker appearance and may have reduced drag. The SF5 also had enlarged vertical tail surfaces. The SF5 was somewhat lighter than the SF2; in light of subsequent experience this lighter structure may have compromised its strength. The final production batch had revised wing bracing to address shortcomings in the original design.

A total of 91 SF5 floatplanes were built in the following series, with minor differences between them. The LVG machines omitted the wireless transmitter for their role as trainers. The first two LVG series were delivered from summer 1917 through the end of the year.

Marine Nr: 968–987 [20 airplanes]
Marine Nr: 1017–1036 (LVG) [20 airplanes]
Marine Nr: 1214–1223 (LVG) [10 airplanes]
Marine Nr: 1224–1233, [10 airplanes]
Marine Nr: 1352–1371 [20 airplanes]
Marine Nr: 1459–1468 (LVG) [10 airplanes]
Marine Nr: 1514 [1 airplane]

The first production batch by Sablatnig was delivered between January and May 1917. The SF5 was not popular with its crews, who had probably expected an improvement over the earlier SF2 but were disappointed. The SF5 had poor speed and climb rate; moreover, it was not especially stable. Some crews asserted that its rated cruising speed of 110 km/h (68 mph) was instead its maximum speed, and "Lame Crow" was reportedly its nickname amongst the crewmen who flew it.

Above: SF5 Marine Number 1019, from the second production batch built by LVG, having a bad day. The standard late-war naval camouflage is clearly seen on the upper surfaces of the wings and floats.

Below: SF5 Marine Number 1021, from the second production batch built by LVG, shares duty as a photo background with Friedrichshafen FF33L Marine Number 941 at left. The SF5 wears the standard late-war naval camouflage.

Above: SF5 Marine Number 1230, from the fourth production batch, carried the standard late-war naval camouflage. The large radiator suspended from the upper wing created a lot of drag, as did the legion of struts and bracing wires. The additional pair of struts bracing the outer wingtips on the upper wing replaced the pylon struts over the outer bay of interplane struts used on the SF2, giving the SF5 a cleaner appearance.

Below: This SF5 is one of two captured and used by the Russians. Russian cockades are visible under the upper wings.

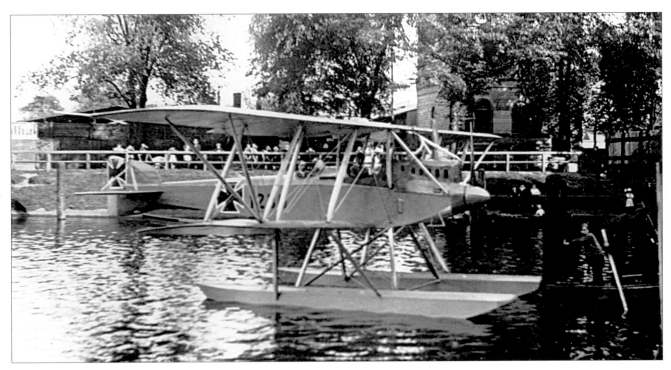

Above: SF5 Marine Number 1230 from the fourth production batch.

Below: This SF5 Marine Number 1361 from the sixth production batch illustrates its revised wing and bracing. The wings were now more equal in span than earlier SF5 production aircraft and the cowling more completely encloses the engine. The tail has also been revised and the rudder has been enlarged with an aerodynamic balance.

Sablatnig SF5 SVK Drawing

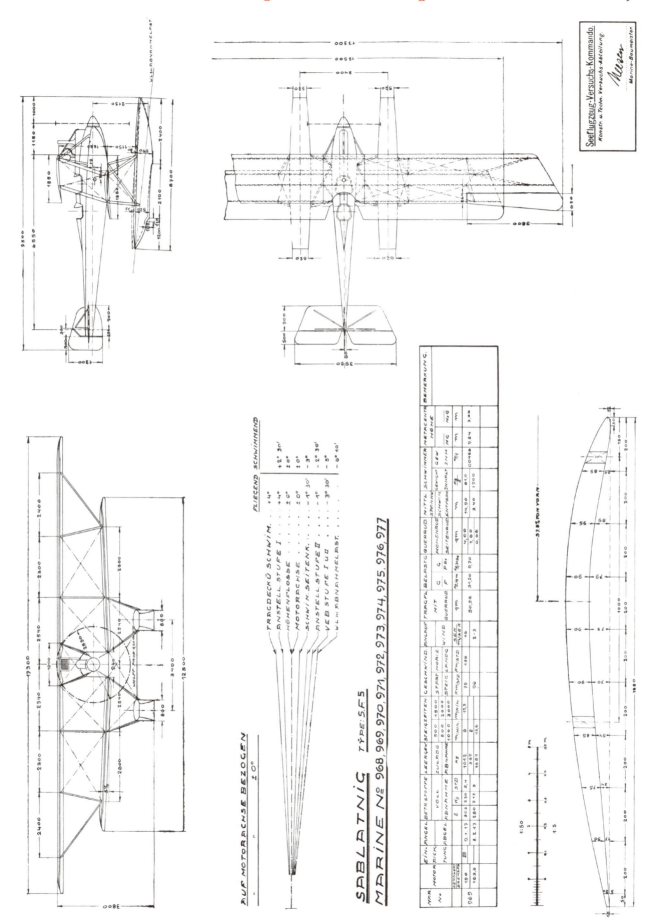

Sablatnig SF5 SVK Drawing

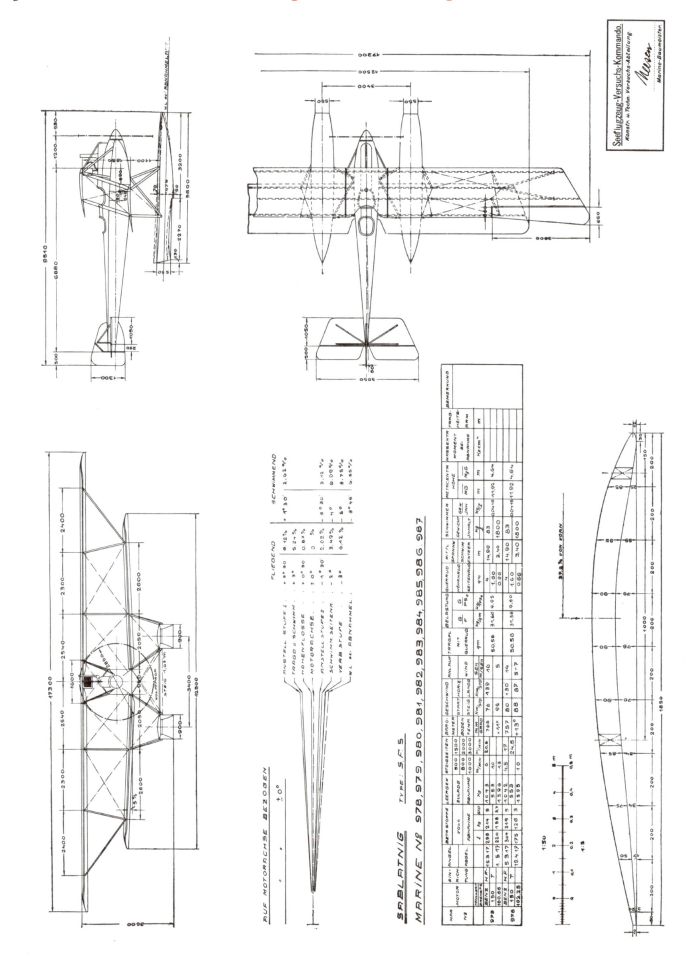

Sablatnig SF5 SVK Drawing

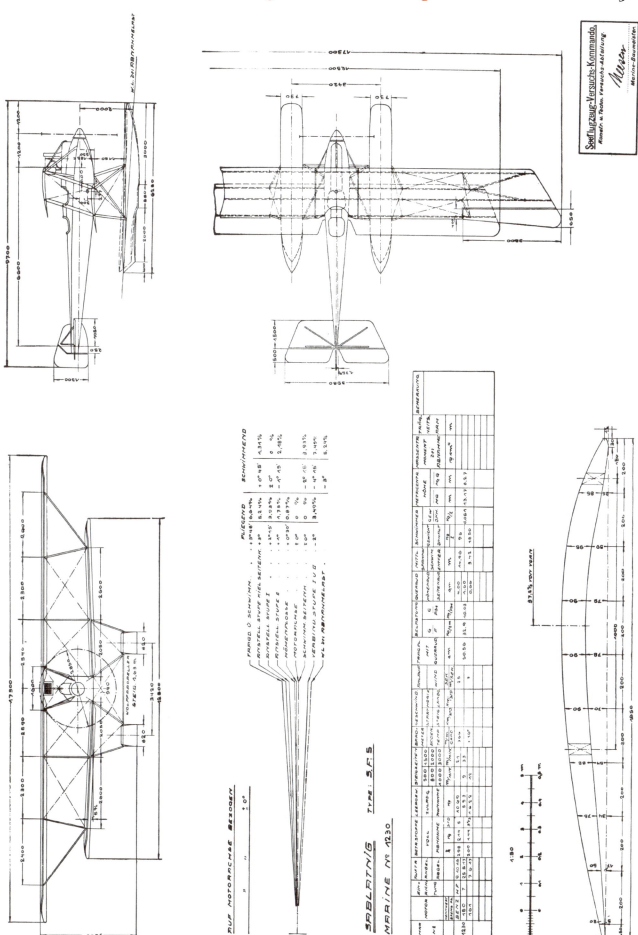

Sablatnig SF5 SVK Drawing

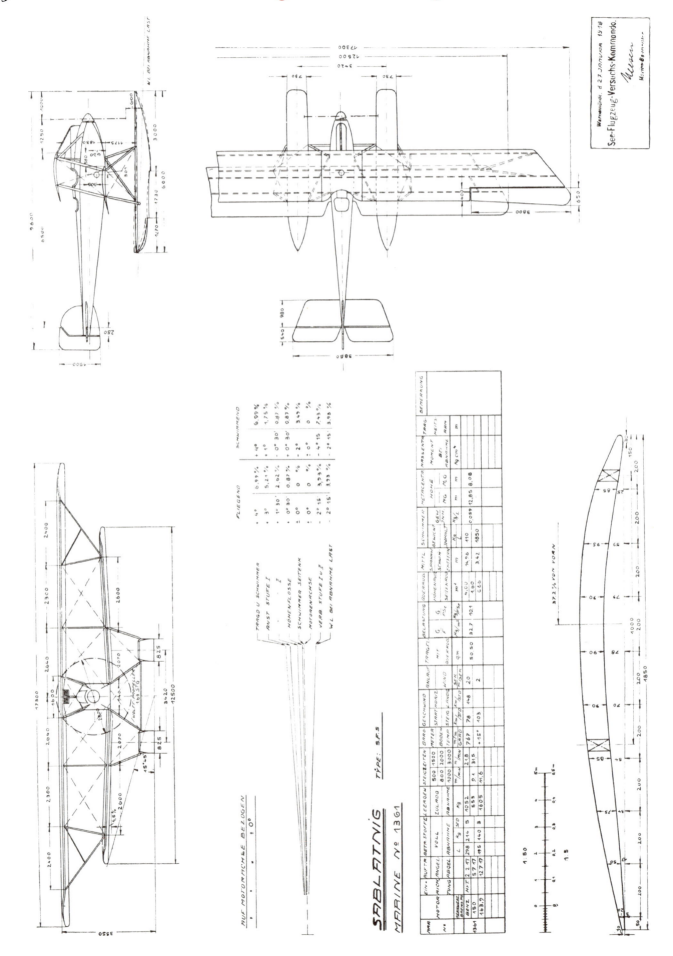

Sablatnig SF5 SVK Drawing

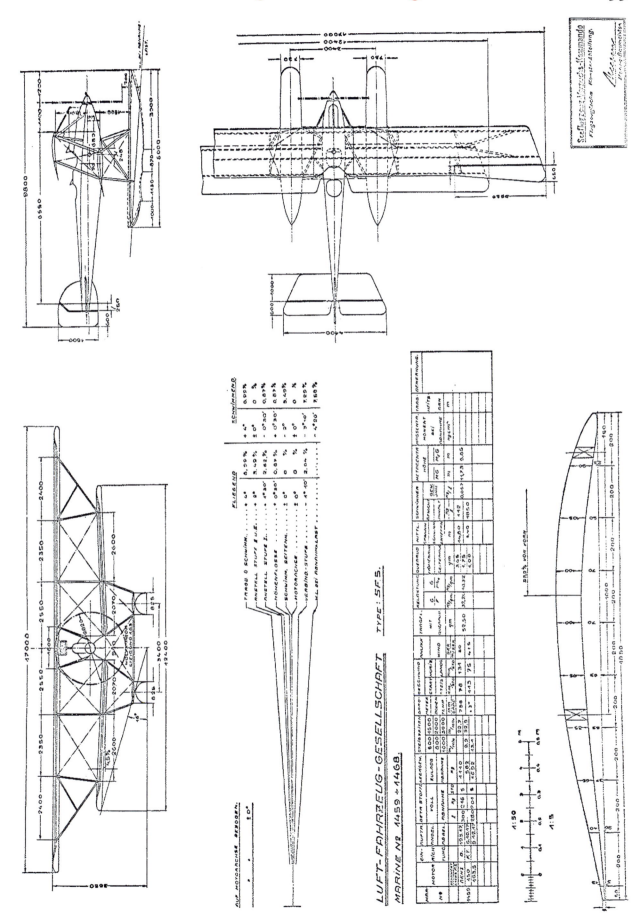

Sablatnig SF6 / B.I

Above & Below: The SF6 was a landplane derivative of the SF5 intended for training. Interestingly, standard late-war naval camouflage appears to have been applied, likely due to Sablatnig's familiarity with it and supply of printed fabric.

The SF6 land plane, a derivative of the SF5 intended for training, is included here for continuity of the SF series and because it is not in the N-type book. Also known as the Sablatnig B.I, it was powered by the same 150 hp Benz Bz.III as the SF5. As far as is known, only one aircraft was built.

Above: The SF6 was a landplane derivative of the SF5 intended for training. Apparently only one was built.

Sablatnig SF5 Marine #968

Sablatnig SF7

Above: The SF7 was a massive aircraft, looking more like a typical reconnaissance floatplane than a two-seat fighter. The competing W19 proved to be superior and was placed in production. Marine #1475 was the first of three prototypes.

Developed from the earlier SF3 prototype, the SF7 was designed as a two-seat floatplane fighter in the C2MG category. Powered by the 240 hp Maybach Mb.IV, three aircraft, Marine Numbers 1475–1477, were ordered, although #1475 may have been the only one built. The SF7 featured I-struts for better streamlining and the inboard wing bay was strut-braced and wireless. The SF7 was a large aircraft that looked more like a conventional two-seat reconnaissance airplane than a fighter. However, it had ailerons on all wings assisted by servo tabs to reduce the pilot's control forces; those may have given it better maneuverability than expected for its size. The SF7 was faster than the competing Brandenburg W19 biplane that used the same engine, but slower than the W33 monoplane that also used the same engine. However, the Brandenburg W12 and its derivatives were robust, maneuverable aircraft with excellent combat records, and the SF7 was not placed in production.

Below: The Sablatnig SF7 was in the W19 class and was powered by the same engine. The Friedrichshafen FF48 was the third competitor for this requirement. The I-struts appear to interfere with the crews' lateral field of view.

Sablatnig SF8

Above & Below: The SF8, a three-bay development of the SF5, was a dual-control floatplane trainer with 150 hp Benz Bz.III engine. Marine Number 2021 is shown below and is probably the aircraft above as well.

A development of the earlier SF5 with revised, three-bay wing bracing, the SF8 was a dual-control floatplane trainer powered by the same 150 hp Benz Bz.III used in the SF5. Three prototypes, Marine Numbers 2020–2022 were built, and a production batch of 30, Marine Numbers 6001–6030, was ordered. Because SVK delivery data stops in June 1918, it is not known if all were delivered.

To simplify naval aircraft identification, on 16 November 1917 different manufacturers were assigned their own blocks of serial numbers. Sablatnig was assigned the Marine Number block 6001–6500. Production was planned as below:

Marine Numbers 6001–6030   SF8
Marine Numbers 6031–6050   FF49C(Sab)
Marine Numbers 6051–6500   —

Above: SF8 Marine Number 2021 floatplane trainer; ailerons on all wings were connected by an actuating strut.

Sablatnig SF5 Marine #1021

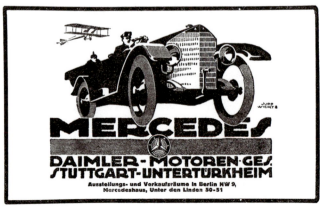

Above & Below: Views of late-production SF8 Marine Number 6006 floatplane trainer at play, perhaps postwar.

Sablatnig SF8 SVK Drawing

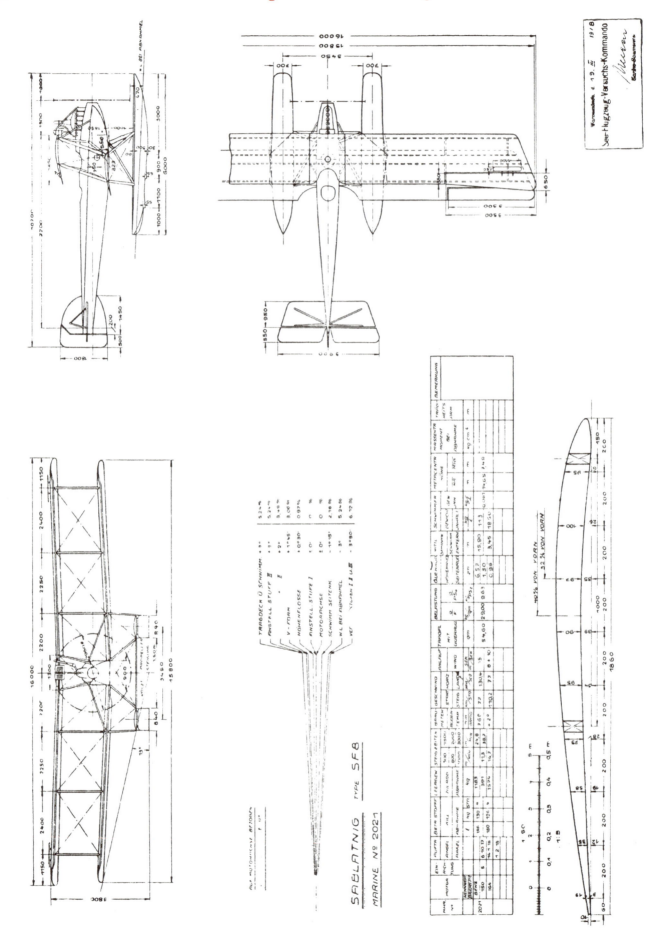

41

Sablatnig SF5 Marine #1230

Sablatnig SF5 Marine #1361

Sablatnig SF5 captured and flown by the IRAS

Kaiserliche Werften Floatplanes

The *Kaiserliche Werften* were government-owned shipyards that were responsible for the construction and repair of warships of the Prussian Navy, later the Imperial Navy, from 1871 to 1920. There were three of these shipyards, one each at Danzig, Kiel, and Wilhelmshaven.

Surprisingly, despite the facts that floatplanes were available from a number of established manufacturers and the *Kaiserliche Werften* were shipyards with no aviation experience, the three *Kaiserliche Werften* designed and built a small number of floatplanes. The rationale given was that the output of the major German seaplane manufacturers was taken up producing machines for front-line service. As a consequence, the only machines available for training purposes were those that were obsolete or which had been damaged and rebuilt, and the *Kaiserliche Werften* designed modern floatplane trainers for the navy.

No doubt there is an element of truth to that, but it would have been far easier, faster, and cheaper to purchase trainers from an established manufacturing company or build them under license than to invest the time and resources to design new aircraft that were built in ones and twos by shipyards.

To justify the extra trouble and expense to build what were essentially prototype aircraft, someone in authority at the *Kaiserliche Werften* must have wanted to not only build but design floatplanes. The fact that the first orders for floatplanes built by the *Kaiserliche Werften* were placed in late 1914, within a few months of the war's beginning, indicates that the authorities at the *Kaiserliche Werften* did not want to wait for production trainers from the established manufacturers and preferred to design and build their own. Furthermore, later in the war the *Kaiserliche Werften* designed and built several armed two-seat floatplanes intended for combat, yet another indication that their primary motivation was to design their own aircraft, not just build them.

Another interesting aspect of the floatplanes built by the *Kaiserliche Werften* is that the same design was built by more than one *Kaiserliche Werft*. Together with some design similarity of aircraft built by the different *Kaiserliche Werften*, this points strongly to the possibility that there was one central design team that was responsible for most, perhaps all, of these aircraft.

Please refer to the SVK Table of *Kaiserliche Werften* Seaplane Orders and Deliveries below for details of the order and delivery quantities and dates.

KW Type 401

The *Kaiserliche Werften* Type 401 was built at *Kaiserliche Werft* Wilhelmshaven (three aircraft, Marine numbers 401–403) and KW Danzig (two aircraft, Marine numbers 404 & 405). These were unarmed, two-seat trainers powered by 100 hp Mercedes D.I engines. They were conventional, three-bay biplanes with the wire-braced, fabric-covered wood structures typical for the time. Technical details are not known.

SVK Table of Kaiserliche Werften Seaplane Orders and Deliveries

Order Number	Type	Marine Numbers	Builder	Class & Engine	1914 J	F	M	A	M	J	J	A	S	O	N	D	1915 J	F	M	A	M	J	J	A	S	O	N	D
1	401	401/403	W'haven	B 100M												3			1	1		1						
2	401	404/405	Danzig	B 100M												2												
3	462	461/462	W'haven	B 150B																						2		
4	462	463/466	Kiel	B 150B																						2		
5	470	467/470	Danzig	B 150B																						4		
6	945	945	W'haven	C 150B																								
7	947	947	W'haven	CHFT 220M																								
8	1106	1105/1106	Danzig	B 150B																								
9	1650	1650	Danzig	CHFT 220M																								
				Orders that Month												5									8			
				Deliveries that Month															1	1		1						
				Orders that Year												5												8
Orders: 3	Deliveries: 2			**Deliveries that Year**												0												3

Right & Below: Two views of Marine #401, built by KW Wilhelmshaven, operating with the fleet. The KW Type 401 was a conventional, three-bay biplane with 100 hp Mercedes D.I engine. KW aircraft built at Danzig were supplied to the nearby naval base at Putzig for training. In the photo below the upper left wingtip has been damaged, preventing the aircraft from flying back to base. The aircraft built at Wilhemshaven were designated W1, W2, and W3 in Marine number order.

	1916												1917												1918											
	J	F	M	A	M	J	J	A	S	O	N	D	J	F	M	A	M	J	J	A	S	O	N	D	J	F	M	A	M	J	J	A	S	O	N	D
																		1		1																
								1														1													
																		1																		
							2						1	1																						
						①							1			1																				
						①																		On 17 December back to the dockyard												
							②																	1105 back to the dockyard on 7 December												
																		①																		
				1	1		2						1																							
						3								1	2			2				1														
							4																1													
						3																	6													

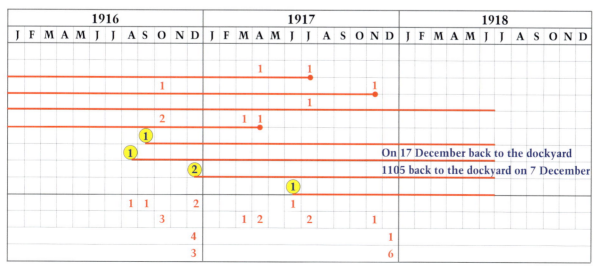

Left: Marine #401 under construction at *Kaiserliche Werft* Wilhelmshaven. The design and construction were typical for the era.

Facing Page, Top: Marine #401 at *Kaiserliche Werft* Wilhelmshaven after being dropped from a crane.

Facing Page, Bottom: Marine #403 at *Kaiserliche Werft* Wilhelmshaven was refined from the earlier #401; it had a more streamlined nose and full-length fuselage decking. The *Kaiserliche Werft* aircraft were essentially prototypes and it is possible that no two were identical.

Left: Marine #401 at *Kaiserliche Werft* Wilhelmshaven after being dropped from a crane. It is not known if the aircraft was repaired after this accident.

Above: Marine #403 ready for flight with naval pennants trailing the lower wings.

Above: Marine Numbers 404 and 405 were built by *Kaiserliche Werft* Danzig to the basic design of the *Kaiserliche Werft* Type 401 although there were likely detail differences between the Wilhelmshaven-built and Danzig-built aircraft.

Kaiserliche Werften Type 462

Above: Marine #461 on a beaching dolly; its interesting cabane and interplane bracing differed significantly from that used by Marine #462 and #463, which followed the SVK drawing of the type. It was built by *Kaiserliche Werft* Wilhelmshaven.

Below: Marine Number 462 was one of two built by *Kaiserliche Werft* Wilhelmshaven; the four aircraft built by *Kaiserliche Werft* Kiel to similar design differed in detail.

Above & Below: Marine #462 appears to be similar in configuration as #463 but does not carry the standard late-war naval camouflage worn by #463. The radiator is offset to starboard on #461–463, whereas the SVK drawing shows the radiator on the centerline. The profile of the rudder and fin on #462 shown in the photo on page 47 is similar to #461 but differs significantly from #463, and none look like the SVK drawing.

The *Kaiserliche Werften* Type 462 was built at *Kaiserliche Werft* Wilhelmshaven (two aircraft, Marine numbers 461 & 462) and KW Kiel (four aircraft, Marine numbers 463–466). These were unarmed, two-seat trainers powered by 150 hp Benz Bz.III engines. They were conventional, two-bay biplanes with the wire-braced, fabric-covered wood structures typical for the time but cleanly designed. The photos show that the interplane bracing of #461 differed notably from that used on #462 and #463.

Kaiserliche Werften Type 462 Specifications

Engine:	150 hp Benz Bz.III	
Wing:	Span	14.9 m
	Wing Area	52 m²
General:	Length	9.5 m
	Height	3.9 m
	Empty Weight	1,016 kg
	Loaded Weight	1,585 kg
Maximum Speed:		143 km/h
Climb to 1,000 m:		10.7 min.

Above & Below: Marine #463, carrying standard late-war naval camouflage, shows its clean lines for a float biplane. The four Kiel-built aircraft were delivered to the naval air station at Kiel-Holtenau. The type was also known as W10.

Kaiserliche Werft Wilhelmshaven 462 SVK Drawing

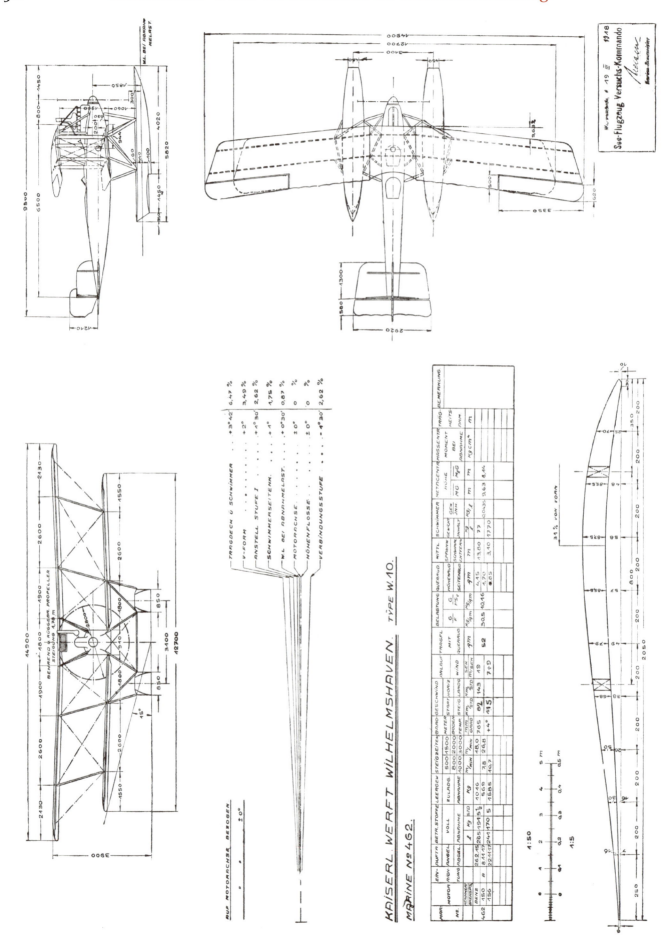

Kaiserliche Werften Type 467

The *Kaiserliche Werften* Type 467 was built at *Kaiserliche Werft* Danzig (four aircraft, Marine numbers 467–470). These were unarmed, two-seat trainers powered by 150 hp Benz Bz.III engines. They were conventional, two-bay biplanes with the wire-braced, fabric-covered wood structures typical for the time. These Danzig-built floatplanes were supplied to the nearby naval base at Putzig.

Kaiserliche Werften Type 467 Specifications		
Engine:	150 hp Benz Bz.III	
Wing:	Span	15.68 m
	Wing Area	52 m²
General:	Length	9.1 m
	Height	3.67 m
	Empty Weight	1,063 kg
	Loaded Weight	1,632 kg
Maximum Speed:		128 km/h
Climb to 1,000 m:		15.0 min.

Kaiserliche Werften Type 945

The only *Kaiserliche Werften* Type 945 was built at *Kaiserliche Werft* Wilhelmshaven. This was an armed, two-seat fighter powered by a 150 hp Benz Bz.III engine; other details have not survived. The aircraft was clearly inspired by the Brandenburg W12. The photograph proves it was completed but SVK records are missing for the period after June 1918 so the actual delivery date is unknown but after June 1918. By this time the fast Brandenburg monoplane W29 and W33 were in combat, so the biplane Type 945, which was undoubtedly slower, had no prospects of quantity production.

Above: The KW Type 945 was a two-seat floatplane fighter category C2MG, designated 'W9' by the *Kaiserliche Werften*.

Kaiserliche Werften Type 947

The only *Kaiserliche Werften* Type 947 was built at *Kaiserliche Werft* Wilhelmshaven. The Type 947 was an armed, two-seat reconnaissance floatplane, naval category CHFT, powered by a 220 hp Mercedes D.IV straight-eight engine.

Kaiserliche Werften Type 947 Specifications		
Engine:	220 hp Mercedes D.IV	
Wing:	Span	15.95 m
General:	Length	13.0 m
	Height	4.00 m

Above: The KW Type 947 was a two-seat reconnaissance aircraft in category CHFT.

Category CHFT meant the aircraft had a flexible gun for the observer and was fitted with a wireless transmitter and receiver.

The Type 947 was ordered in August 1916. Production of the six-cylinder, 260 hp Mercedes D.IVa was in its very early stages so the most powerful engine available was the eight-cylinder, 220 hp Mercedes D.IV. A four-blade propeller was used instead of a two-blade propeller so the individual blades would be shorter, providing good clearance from the water while enabling shorter struts for the floats.

The Type 947 was evaluated by the *SVK* at Warnemünde but was not accepted and was returned to the manufacturer on December 17, 1917. *SVK* records stop at the end of June 1918, so the final disposition of the Type 947 is unknown.

Kaiserliche Werften Type 1105

Two *Kaiserliche Werften* Type 1105 floatplanes, Marine numbers 1105 and 1106, were built at *Kaiserliche Werft* Danzig. In the listing of German seaplane numbers, Marine number 1105 is described as a B/S class (unarmed trainer) and 1106 is listed as a C2MG/S, an armed two-seater intended for training. Both were powered by a 150 hp Benz Bz.III.

Technically, the most interesting aspect of this type is that no bracing wires were needed in the wing cellule due to the strut design.

The two floatplanes were ordered in December 1916 and no delivery date is given, although #1105 is shown as returned to the dockyard on December 7, 1917. At least #1105 was completed and evaluated by the *SVK* but was not accepted and was returned to the manufacturer. Its ultimate status and that of #1106 are not known due to incomplete *SVK* records.

Kaiserliche Werften Type 1105 Specifications		
Engine:	150 hp Benz Bz.III	
Wing:	Span	14.1 m
General:	Length	8.85 m
	Height	3.73 m

Above & Below: The KW Type 1105 was a two-seat trainer, category B/S. Evaluated by the *SVK* in 1917, it was not accepted and was returned to the manufacturer on December 7, 1917. The engine was a 150 hp Benz Bz.III.

Kaiserliche Werft Danzig 470 SVK Drawing

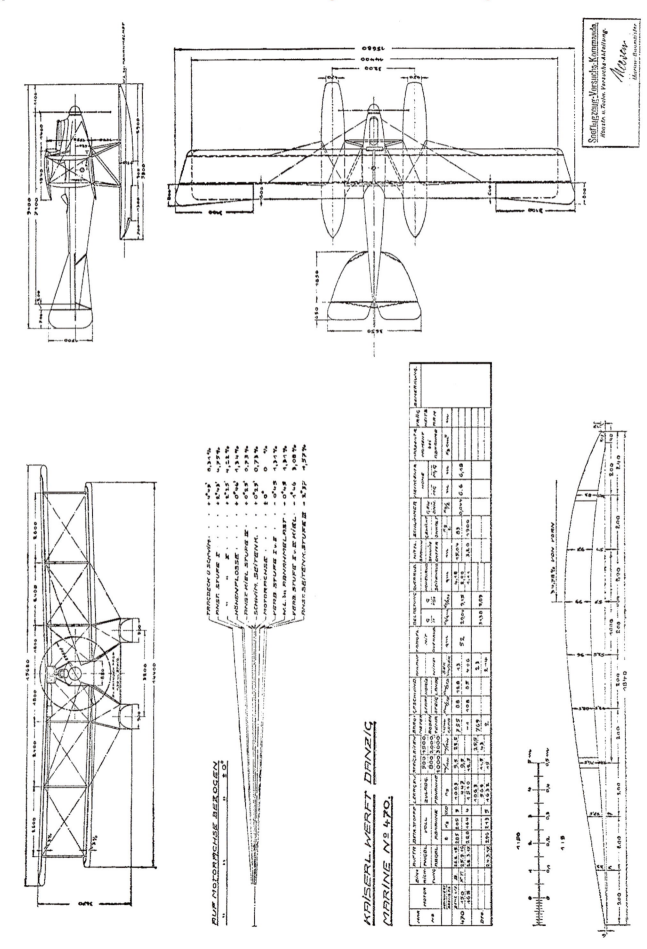

Kaiserliche Werften Type 1650

The sole *Kaiserliche Werften* Type 1650 floatplane was built at KW Danzig. Ordered in June 1917, it was the last floatplane ordered from the *Kaiserliche Werften*. The listing of German seaplane numbers shows it as a category CHFT, meaning the aircraft had a flexible gun for the observer and was fitted with a wireless transmitter and receiver.

By the time the Type 1650 was ordered, the 260 hp Mercedes D.IVa six-cylinder engine was well-established in production. Despite that, the Type 1650 was powered by a straight-eight 220 hp Mercedes D.IV that was then obsolescent and out of production. The straight-eight was not only less powerful than its six-cylinder replacement, it was also less reliable. Crankshaft failures of the long engine were a problem, although much less so in single-engine aircraft than in multi-engine aircraft.

The most likely reason the older, less reliable engine was used is that it was available; most of the few production aircraft using this engine had been retired from the front by the time the Type 1650 was ordered, releasing some engines for other use. The more powerful and reliable Mercedes D.IVa engine that replaced the straight-eight in production was in great demand for G-type bombers and the more powerful C-type reconnaissance airplanes. The Type 1650 had a much lower priority than these vital types and so had to make due with the older engine.

The *SVK* records are incomplete so the delivery date of the Type 1650; indeed whether it was completed or delivered at all is not known. The list of German seaplane number notes that it was cancelled and there is no *SVK* drawing of the type.

Kaiserliche Werften Type 1650 Specifications	
Engine:	220 hp Mercedes D.IV

Above: A Sablatnig N.I provides the background for a group portrait at *Flugstation* Barde. For full coverage of the Sablatnig N.I and the other Sablatnig land planes, see Volume 3 of this series, *Nachtflugzeug! German N-Types of WWI*.

Lübeck-Travemünde Floatplanes

Flugzeugwerft Lubeck-Travemunde G.m.b.H. was founded in May 1914 at Travemunde Privall for the construction of seaplanes. It was a subsidiary of Deutsche Flugzeug-Werke (DFW), best known for its DFW C.V two-seat reconnaissance airplane.

Lubeck-Travemunde produced four floatplane designs and delivered at least 18 seaplanes to the German Navy; SVK records after June 1918 are not complete. Only a handfull of Lubeck-Travemunde F2 and F4 floatplanes flew on operations.

Lübeck-Travemünde F1

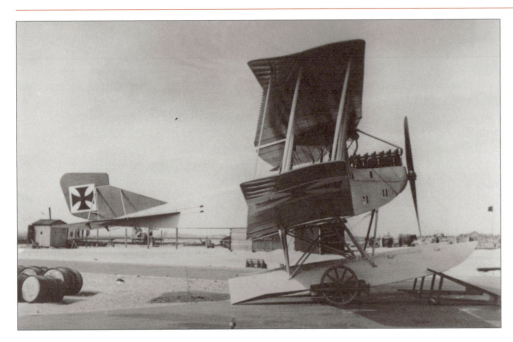

Left: The F1 was the first Lübeck-Travemünde design built. Three aircraft were ordered by only the first, Marine Number 282 seen here, was built; the other two were cancelled. The F1 was a large aircraft for only 160 hp provided by its Mercedes D.III engine and that may have been a key reason only one was built.

The first Lübeck-Travemündes design was the F1, a large, unarmed, four-bay two-seat reconnaissance floatplant, of which three examples were ordered, Marine Numbers 282–284. It was powered by a 160 hp Mercedes D.III engine. The first flight of the F1 was in June 1914. From the SVK records (see below) only Marine Number 282 was built and the other two were cancelled.

SVK Table of Lübeck-Travemünde Seaplane Orders and Deliveries

Order Number	Type	Marine Numbers	Design	Class & Engine		1914												1915											
					J	F	M	A	M	J	J	A	S	O	N	D	J	F	M	A	M	J	J	A	S	O	N	D
1	282	282/283	F1	B 160M												2													
2	282	284	F1	B 160M																									
3	677	677	F2	CFT 220M																									
4	844	844	F3	E 150B																									
5	1147/50	1147/1156	F2	CHFT220M																									
6	1971	1971/1973	F4	CHFT 200B																									
7	1150	1974/1979	F2	CHFT220M																									
8	1971	7001/7030	F4	CHFT 200B																									
9	1971	2135	F4	CHFT 200B																									
				Orders that Month												2												
				Deliveries that Month																								
				Orders that Year												2												–
Orders: 3	Deliveries: 2			Deliveries that Year												–												–

Lübeck-Travemünde F2

Above: Marine Number 677 was the prototype Lübeck-Travemünde F2. The F2 was an enlarged, more powerful F1 powered by the rare 220 hp Mercedes D.IV straight-eight engine. The prototype F2 was a four-bay machine; subsequent F2 floatplanes were three-bay designs. The F2 was a type CHFT, the German naval designation for an armed, two-seat reconnaissance aircraft equipped with a wireless transmitter and receiver. The F2 had a single flexible gun for the observer but no gun for the pilot.

Above, Below, & Top of Facing Page: Additional views of Marine Number 677, the prototype Lübeck-Travemünde F2, emphasizing its massive size and robust construction.

The Lübeck-Travemünde F2 was an improved version of the company's earlier F1 floatplane and was the first armed aircraft built by the company. The F2 was designed as a class CHFT two-seat reconnaissance floatplane powered by a 220 hp Mercedes D.IV straight-eight engine. The observer's rear cockpit was fitted with a wireless transmitter and receiver and a single flexible Parabellum machine gun. There was no gun for the pilot.

The prototype F2, Marine Number 677, was a massive, four-bay biplane. It was ordered in March 1916 and delivered to the SVK in December 1916.

Ten production aircraft, Marine Numbers 1147–1156, were ordered in December 1916. These and all subsequent F2 floatplanes were three-bay biplanes. Deliveries of these floatplanes started in August

1917 and were completed in February 1918. Another order for six floatplanes, Marine Numbers 1974–1979, was placed in October 1917; delivery of these aircraft started in March 1918, immediately after completion of the first production order. Delivery of five of these aircraft is verified by SVK records, which are incomplete after June 1918, making a total of 16 aircraft confirmed delivered of the 17 ordered.

Lübeck-Travemünde F2 #677 Specifications		
Engine:	220 hp Mercedes D.IV	
Wing:	Span	19.0 m
	Wing Area	86 m²
General:	Length	11.32 m
	Height	3.575 m
	Empty Weight	1,540 kg
	Loaded Weight	2,204 kg
Maximum Speed:		136 km/h
Climb to 1,000 m:		10.0 min.

Above: Lübeck-Travemünde F2 Marine Number 1151 of the main production order represents the third variation of the F2 design and is representative of most F2 production floatplanes. Differences from the prototype include revised tail surfaces on a longer fuselage, three-bay wings, and more robust float bracing.

Above: Lübeck-Travemünde F2 Marine Number 1978 was the next to last F2 ordered and was likely delivered in May 1918, the last confirmed delivery of a production F2 floatplane. Like Marine Number 1151 below and the majority of the F2 production aircraft it was built to the final F2 configuration. The robust interplane and float struts are clearly visible in this closeup photograph. The unusual straight-eight Mercedes engine was probably used because the newer and more powerful 260 hp Mercedes D.IVa six-cylinder was in great demand for bombers and Rumpler reconnaissance aircraft.

Below: Lübeck-Travemünde F2 Marine Number 1151 of the main production order.

Lübeck-Travemünde F2 SVK Drawing

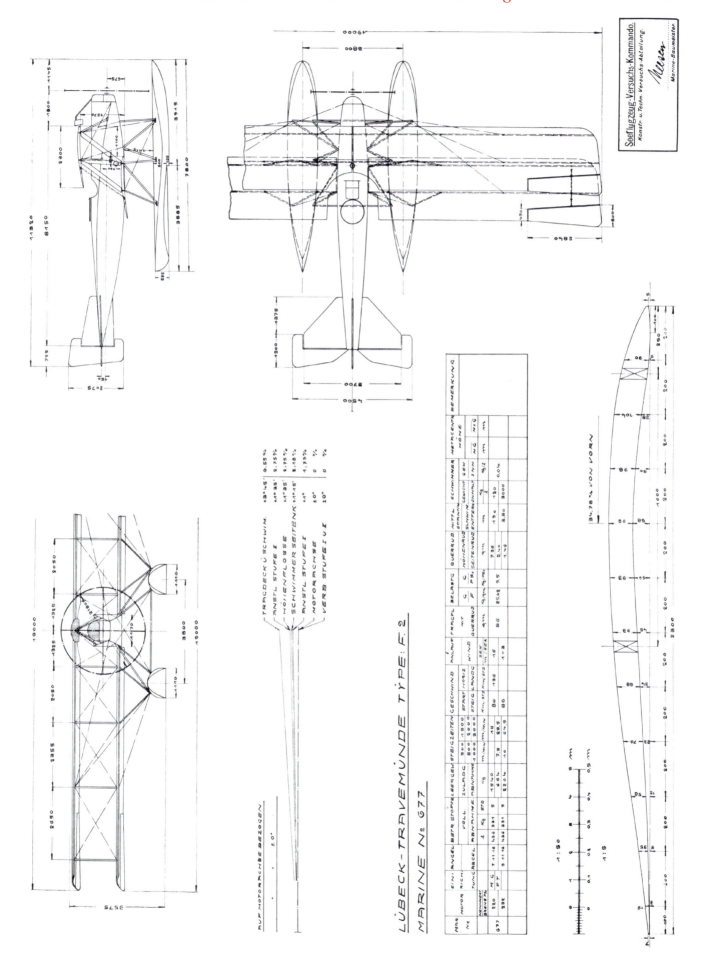

Lübeck-Travemünde F2 SVK Drawing

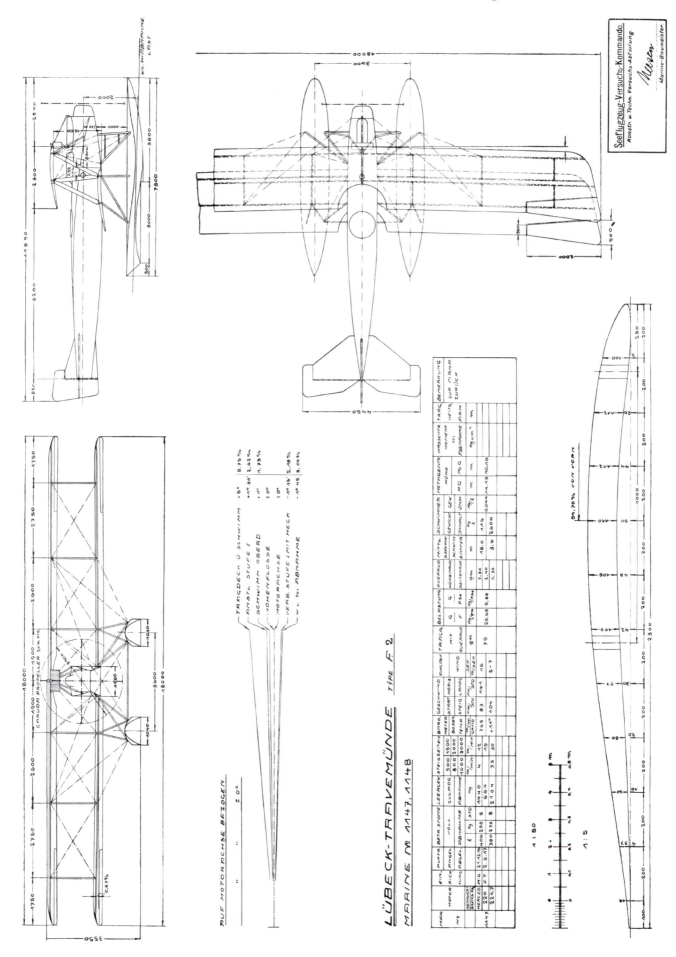

Lübeck-Travemünde F2 SVK Drawing

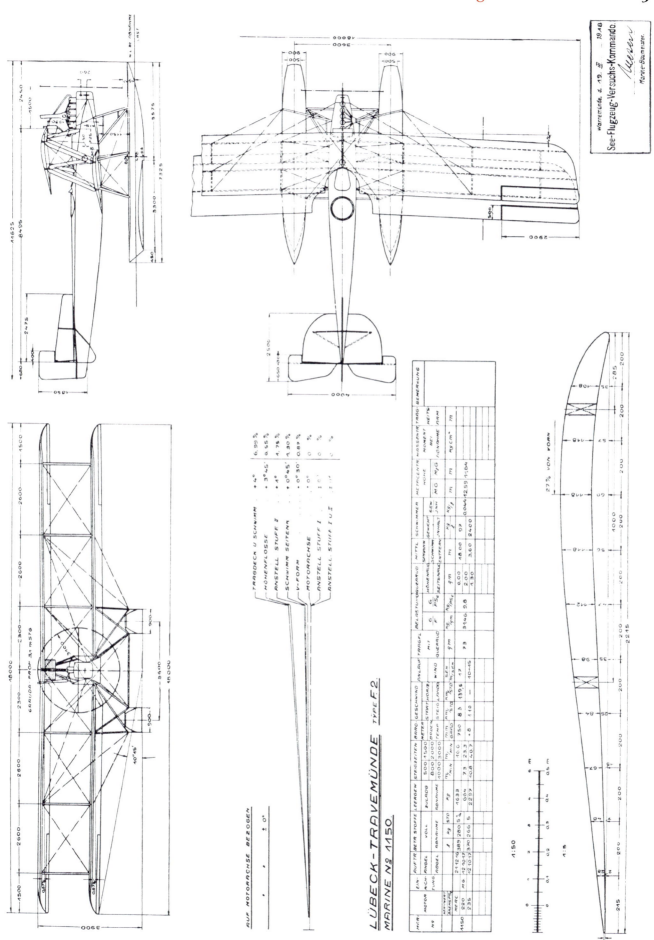

Lübeck-Travemünde F3

Above: The Lübeck-Travemünde F3 single-seat fighter, Marine Number 844, was powered by a 150 hp Benz Bz.III.

Little is known about the Lübeck-Travemünde F3 single-seat floatplane fighter and only the above photograph has been discovered. Ordered in July 1916 when the German Navy was determined to procure single-seat floatplane fighters to protect its naval air stations, the F3 was powered by the 150 hp Benz Bz.III, the same engine used in many of its competitors.

The F3 seems to have suffered prolonged development, but because the *SVK* did not accept the F3 its completion and testing dates are not known. By the end of 1917 the Brandenburg W12 two-seat floatplane fighter had proven itself in combat.

Lübeck-Travemünde F3 Specifications	
Engine:	150 hp Benz Bz.III

The Brandenburg W12 clearly demonstrated the improved combat effectiveness of a good two-seat seaplane fighter compared to single-seaters, and the German Navy was no longer interested in single-seat floatplane fighters, even proven designs like the Albatros W4 and the Rumpler 6B types. The *SVK* declined acceptance of the F3 on March 23, 1918 and only the single F3 was built. Other than the powerplant, no other technical data has survived.

Lübeck-Travemünde F4

Above: The Lübeck-Travemünde F4, Marine Number 1971, was the prototype F4.

The Lübeck-Travemünde F4 was the final Lübeck-Travemünde design to be ordered and built. Like the earlier F2, it was a class CHFT armed, two-seat reconnaissance floatplane equipped with wireless transmitter and receiver. Like the F2, only the observer had a gun.

The F4 was a somewhat smaller, lighter aircraft than the preceding F2 and had the less powerful but more reliable 200 hp Benz Bz.IV six-cylinder engine. The first batch of three prototype aircraft, Marine Numbers 1971–1973, was ordered in October 1917 in parallel with the last F2 production batch. Marine Number 1971, the first prototype, was delivered in March 1918. Marine Number 1972 was destroyed during structural testing and Marine Number 2135 was ordered to replace it.

Thirty F4 floatplanes, Marine Numbers 7001–7030, were ordered in February 1918 before delivery of the prototype. Because *SVK* delivery information past June 1918 is missing it is not known how exactly many of these were delivered. However, the photograph of Marine Number 7022 indicates that most, and perhaps all, of this batch were delivered.

Lübeck-Travemünde F4 Specifications		
Engine:	200 hp Benz Bz.IV	
Wing:	Span	16.7 m
	Wing Area	57.64 m²
General:	Length	11.3 m
	Height	4.0 m
	Empty Weight	1,366 kg
	Loaded Weight	1,998 kg
Maximum Speed:		138 km/h
Climb to 1,000 m:		9.3 min.

Above & Below: Two more photographs of the prototype Lübeck-Travemünde F4, Marine Number 1971. In these views the three-color marine camouflage fabric on the upper surfaces of the wings is just visible. A somewhat smaller, lighter development of the earlier F2 designed for the same role, the F4 had the less powerful but more reliable 200 hp Benz Bz.IV engine. Like the production models of the F2, the F4 was a robust, three-bay biplane.

Left: This photo was labelled "Lübeck-Travemünde Trainer" and the aircraft does have some family resemblance to the Lübeck-Travemünde F4, although it is a smaller, less powerful two-bay land plane. The late-style markings confirm it was photographed in 1918. Naval landplane number 102 is listed as an LVG, although the type is not given, but these numbers were applied early in the war, not in 1918. It is more likely that the full serial number on this B-type aircraft is 102/18, but that is speculation as only the "102" can be seen. Further information is welcome.

Above: This photograph of Lübeck-Travemünde F4 Marine Number 7022 proves that some, and perhaps all, of the aircraft in the production batch of 30 were delivered.

Below: An F4 in postwar Norwegian service. Two F4 aircraft served in the Norwegian Navy; they had more powerful Benz engines of 220–260 hp, enabling them to carry a 500-kg torpedo during torpedo experiments as shown here.

Lübeck-Travemünde F4 SVK Drawing

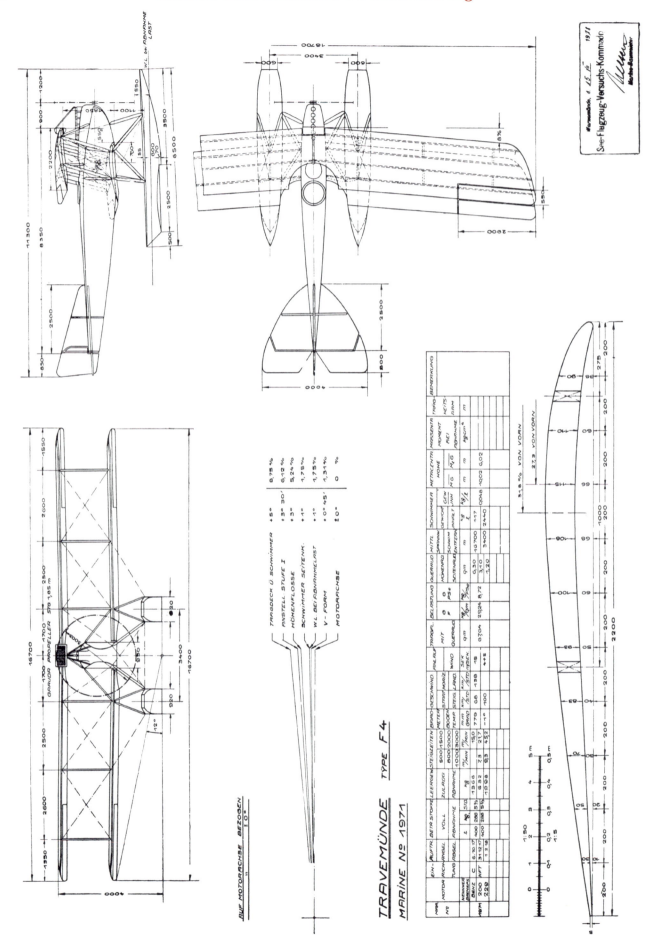

Lübeck-Travemünde F2 #677

Lübeck-Travemünde F2 #1151

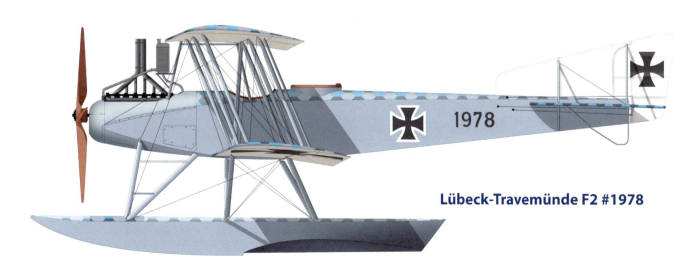

Lübeck-Travemünde F2 #1978

Luft Torpedo Gesellschaft Floatplanes

The Luft Torpedo Gesellschaft (LTG) was established in Berlin in March 1915 to develop an air-launched torpedo. It expanded into sub-contract work building aircraft assemblies and on 8 February 1917 received an order for three prototypes of a floatplane fighter of its own design.

SVK Table of Luft Torpedo Gesellschaft Seaplane Orders and Deliveries

Order Number	Type	Marine Numbers	Design	Class & Engine	1917												1918											
					J	F	M	A	M	J	J	A	S	O	N	D	J	F	M	A	M	J	J	A	S	O	N	D
1	1299	1299/1301	5D1	C2MG 150B		③				1		1		1299 Destroyed in structural test 30 July 1917														
2	1518	1518/1520	5D1	C2MG 150B							③	1300 Back to LTG 7 Sep. 1917					1					2					
				Orders that Month	3			3																				
				Deliveries that Month						1		1							1				2					
				Orders that Year										6												–		
Orders: ③	Deliveries: 2			Deliveries that Year										2												3		

LTG FD 1

Above: Marine number 1299 was the first prototype of the FD 1 floatplane fighter. Power was provided by a 150 hp Benz Bz.III engine driving the propeller through a Loeb reduction gear, accounting for the propeller rotating in opposite direction to that of the engine.

The FD 1 was the only aircraft designed and built by the Luft Torpedo Gesellschaft (LTG). Three prototypes of the FD 1 were ordered on 8 February 1917. Marine number 1299, the first prototype, was

Above: Marine number 1299, the first prototype of the FD 1 floatplane fighter, was later destroyed during structural testing. The man holding the tail has been partially edited out of the photo.

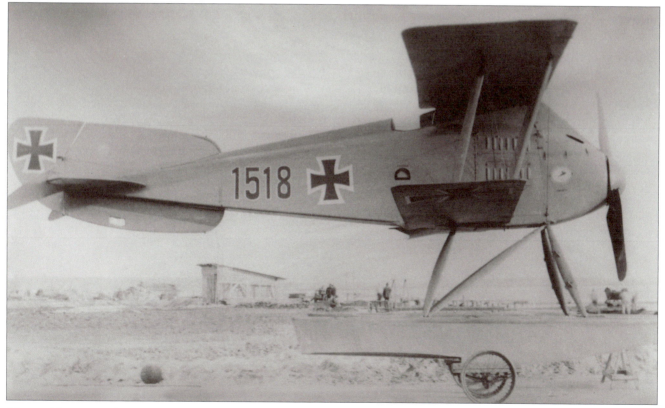

Above: Marine number 1518 was the first aircraft of the redesigned FD 1 fighter batch. The vertical tail surfaces were considerably enlarged compared to the first three prototypes. In addition, the lower wing was modified and the interplane struts were now almost parallel.

Above: FD 1 Marine Number 1518 was the first of three of the re-designed, strengthened FD 1 fighters ordered; the vertical tail was greatly enlarged compared to the first version. Testing was incomplete when the war ended. It is difficult to understand why an experimental single-seat floatplane fighter was being tested this late in the war after the great success of the Brandenburg two-seat floatplane fighters.

LTG FD 1 Specifications		
Engine:	150 hp Benz Bz.III	
Wing:	Span	10.0 m
General:	Length	9.0 m
	Height	3.55 m
	Empty Weight	895 kg
	Loaded Weight	1,165 kg
Maximum Speed:		145 km/h
Climb to 1,000 m:		4.5 min.

delivered in May 1917. Unusually, the FD 1 used a reduction gear to drive the propeller, enabling the engine to be fully enclosed and resulting in the propeller's direction of rotation being opposite to that of the engine. Flight testing revealed poor maneuverability and a second batch of three revised prototypes was ordered. These aircraft had greatly enlarged vertical tail surfaces, indicating longitudinal stability of the initial design also needed improvement.

The second batch of fighters was delivered in 1918, well after the two-seat floatplane fighter had demonstrated its operational superiority over single-seat fighters. Marine number 1299 was destroyed during structural testing and the remaining five aircraft were placed in storage at Hage where the Allies discovered them in December 1918.

FD 1 Marine #1518

LTG FD1 SVK Drawing

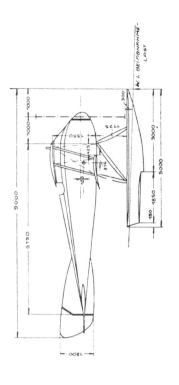
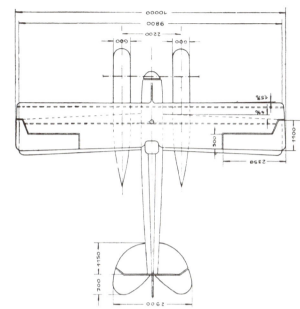

Oertz Flying Boats

Max Oertz was born on 20 April 1871 in the town of Neustadt in Holstein, Germany. He studied nautical design in Berlin at the Royal Institute of Technology at Charlottenburg. He then worked as a designer in Helsinki (then Helsingfors) and St. Petersburg before returning to Berlin in 1895. He then built a yacht for a Berlin banker that was the first to be built completely from aluminum. In 1896 he founded Werft Oertz & Harder together with his friend Hans Harder, and in 1902 bought out his partner. He continued to build exceptional yachts, including one for the Kaiser.

Dr.-Ing. Oertz became interested in aviation and built an observation car that could be lowered from Zeppelins; this enabled the airship to remain invisible above a cloud layer while the observer in the streamlined car would be beneath the clouds with good visibility.

Oertz moved on to designing and building airplanes. He built his first powered aircraft in 1909 and in 1911 he built another with covered fuselage. Oertz then combined his enthusiasms for airplanes and yachts and started building flying boats. Oertz built his first two flying boats before the war, and the Oertz F.B.3 was his first design purchased by the German Navy. He went on to build a small number of flying boats for the Navy before selling his aviation interests to Hansa-Brandenburg in 1917.

Post-war Oertz continued in shipbuilding and design, creating numerous developments and innovations. In 1922 Oertz sold his shipyard and continued on as a freelance designer. Oertz passed away from a heart attack on November 24, 1929, in Hamburg.

The Oertz flying boats were distinctive. From his famous yachts Oertz was known for exceptional workmanship and attention to detail, and his flying boats benefitted from that heritage. In addition, Oertz brought the creativity demonstrated in his yacht designs to his flying boat designs. Oertz's first priority for his flying boats was seaworthiness; to ensure their success the flying boats all had broad-beamed hulls. As a result, his flying boats were so stable that a crewman could walk out half way on the wings without the wing tilting into the water.

To further ensure lateral stability, the engines of the Oertz flying boats were mounted in their hulls at the center of gravity and drove their pusher propellers via drive shafts attached to bevelled gears. The lower wing span was larger than the upper wing span for greater in-flight stability, and to ensure equal strength the interplane struts were slanted inboard at their tops.

Although the F.B.1 had typical wing-tip floats to protect the lower wings, most later Oertz flying boats used small, spring-loaded skis under the outer wing-tips to reduce weight and drag.

Because the German Navy preferred floatplanes for the cold, turbulent northern waters where it operated, only a very few flying boats were ordered from any manufacturer, including Oertz. This meant that Oertz manufactured some Sablatnig floatplanes under license.

Oertz also provided flying boat designs to Friedrichshafen for their FF11 and FF21. In 1917 the Hansa und Brandenburgische Flugzeugwerke purchased the Oertz aviation business and *Dr.-Ing.* Max Oertz returned to the shipping business.

SVK Table of Oertz Flying Boat Orders and Deliveries

Order Number	Type	Marine Numbers	Design	Class & Engine	1914 J F M A M J J A S O N D	1915 J F M A M J J A S O N D
1		46	F.B.3	160M	1	
2		63	W.4	160M	1 1	
3		75	W.4	115Arg		1
4	280	276/280	W.5	C 240Mb	5	
5	281	281	W.6	2x240Mb	1	
6		474/475	W.7	C 150Mb		2
7	1157	1157	W.8	C 240Mb		Gift of the Krupp Compa
				Orders that Month	6	2
				Deliveries that Month	1 1 1	
				Orders that Year	6	2
Orders: 3	Deliveries: 2			**Deliveries that Year**	2	1

Oertz F.B.1

Above: The Oertz F.B.1 was a pre-war design, being built in 1913. The F.B.1 was clearly a boat hull with flying surfaces.

The first Oertz flying boat, creatively designated the F.B.1, was designed and built in the spring of 1913. Powered by a 100 hp Argus As.I driving a pusher propeller, the engine was mounted higher then on later designs. Only one aircraft was built.

The F.B.1 featured streamlined wing-tip floats, slightly-swept back three-bay wings of equal chord, and large tail surfaces mounted high above the water. Like all Oertz flying boats the hull had a very broad beam for stability.

The F.B.1 was unarmed and designed for two crewmen sitting side-by-side. Subsequent Oertz flying boat designs were clearly developed from the F.B.1 based on experience gained from that type.

	1916												1917												1918											
	J	F	M	A	M	J	J	A	S	O	N	D	J	F	M	A	M	J	J	A	S	O	N	D	J	F	M	A	M	J	J	A	S	O	N	D
					1												1	1		1					1											
																			1																	
				1	1																															
ny						1																														
				2	1		1							1	1		1	1							1											
												—												—												
												4												4												

Oertz F.B.2

Above: The Oertz F.B.2 was another pre-war design, being built in 1913. The F.B.2 was a refinement of the F.B.1.

The second Oertz flying boat to be built in 1913 was the F.B.2. Whereas the earlier F.B.1 looked like a speedboat fitted with wings and tail, the F.B.2 was a more refined design. The wooden hull was more streamlined, the tail was mounted above the lengthened hull, and the wing cellule was markedly different. Unlike the F.B.1, the lower wing of the F.B.2 had greater span and chord than the upper wing. The wing-tip floats of the F.B.1 were replaced by spring-loaded boards, which reduced weight and drag. The lower wing tips were enlarged with expansive, swept-back ailerons that curved upward.

The F.B.2 was powered by a 120 hp Argus As.II mounted in a manner similar to the earlier F.B.1, but the engine was lower in the hull. The radiator was mounted in front of the engine and the pusher propeller was driven by a shaft with bevel gears. The crew of two sat side by side in an open cockpit. The sole F.B.2 built was unarmed.

Sablatnig SF2 Marine #795

Oertz F.B.3

Above: Built in early 1914, the Oertz F.B.3 was purchased by the Navy and assigned Marine Number 46. Here it is shown after national markings and Marine Number have been applied between Friedrichshafen FF29 floatplanes #208 and #204. Marine Number 407 in the background is another Friedrichshafen FF29.

In early 1914 the F.B.3 was built; it closely followed the F.B.2 design but had a hull step to improve take-off performance. A much more powerful engine, a 160 hp Mercedes D.III, was installed to further improve take-off performance.

The F.B.3 was purchased by the German Navy in July 1914 and given Marine Number 46. It was flown operationally from Zeebrugge and was later transferred to the Baltic. Like earlier Oertz flying boats, only one F.B.3 was built.

Oertz F.B.3 Specifications		
Engine:	160 hp Mercedes D.III	
Wing:	Span	14.30 m
General:	Length	10.0 m

Above: The Oertz F.B.3 shown before national markings and Marine Number were applied.

Oertz-Flugboot

Oertz W4

Above & Below: The Oertz W4 photographed with national markings but before its Marine Number was applied. The excellent workmanship for which Oertz was famous is apparent. Two crewmen sat side by side in an open cockpit.

Two W4 flying boats, which were virtually identical to the F.B.3, were built. The first, Marine Number 63, had a 160 hp Mercedes D.III and was delivered in November 1914. Marine Number 75 had a 115 hp Argus As.II and was delivered in January 1915. Both flying boats served in the Baltic.

Oertz W4 Specifications		
Engine:	160 hp Mercedes D.III or	
	115 hp Argus As.II	
Wing:	Span	14.30 m
General:	Length	10.0 m

Above: The Oertz W4 photographed in front of the Oertz factory.

Below: This photograph published in *Flight* magazine in 1919 shows either the F.B.3 or a W4 with flexible gun mounting.

Oertz W5

Above: Five examples of the Oertz W5, Marine Numbers 276–280, were ordered in August 1914. The W5 was powered by the 240 hp Maybach Mb.IV, giving it much more power than earlier Oertz flying boats.

The Oertz W5 design was similar to the earlier Oertz flying boats but it was larger and more powerful. It featured a three-bay wing and was powered by the 240 hp Maybach Mb.IV. Five boats, Marine Numbers 276–280, were ordered in August 1914, making the W5 the Oertz type constructed in the largest numbers. Production was extremely slow; the first was not delivered until May 1916 and the last was not delivered until February 1918.

Oertz W5 Specifications		
Engine:	240 hp Maybach Mb.IV	
Wing:	Span	18.0 m
	Area	77.0 m²
General:	Length	10.7 m
	Empty Weight	1,580 kg
	Loaded Weight	2,250 kg
Maximum Speed:		120 km/h

Above: Oertz W5 Marine Number 276 on a beaching dolly. The W5 followed the configuration of earlier Oertz designs but was larger and more powerful. Five examples of the W5 were built, making it the Oertz design built in greatest quantity.

Below: Oertz W5 Marine Number 276 getting ready to take off. All available photos of the Oertz W5 are of Marine Number 276.

Oertz W5 SVK Drawing

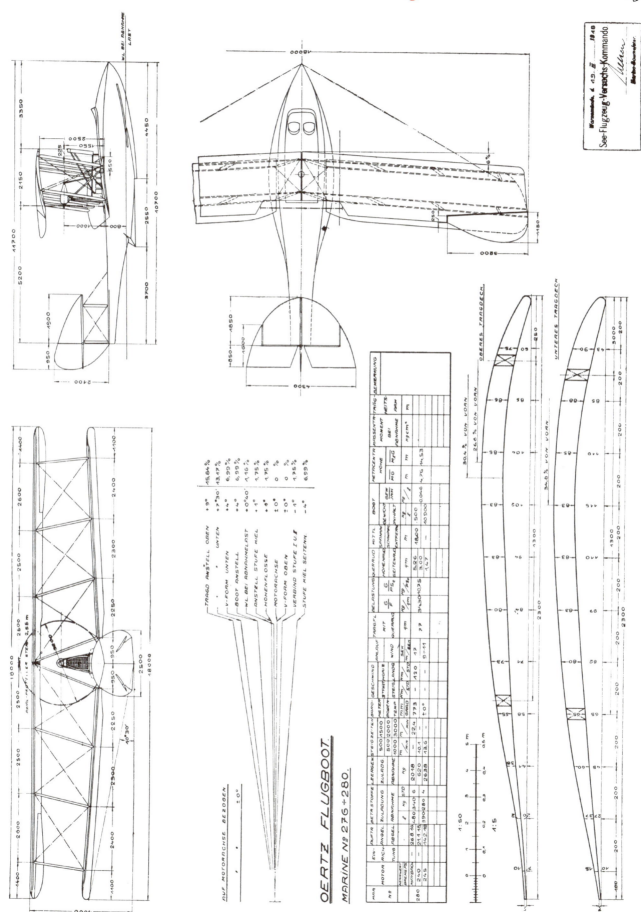

Above: Oertz W5 Marine Number 276; all available photos of the W.5 are of this aircraft, presumably the first aircraft in the series to be completed.

Left: Oertz W8 Marine Number 1157. The W8 was the last Oertz flying boat design and only one was built. Unlike other Oertz designs such as the W5 above, the W8 featured a balanced rudder.

Oertz W6 Flying Schooner

Above & Below: Oertz W6 Marine Number 281, with its tandem biplane design, was the most distinctive Oertz flying boat. Known as the *Flugschoner* (Flying Schooner), it was powered by two 240 hp Maybach Mb.IV engines. In original configuration above, the below photo shows it after auxiliary ailerons were added between the rear pair of wings.

Above & Below: Oertz W6 Marine Number 281, in its original configuration before auxiliary ailerons were added between the rear pair of wings and before wing-tip floats were added.

The Oertz W6, known as the *Flugschoner* (Flying Schooner), was ordered in August 1914 but was not delivered until July 1917. The W6 was easily the most distinctive Oertz flying boat due to its tandem biplane wing design. This configuration was chosen to provide great wing area within a reasonable wing span. The W6 was powered by two 240 hp Maybach Mb.IV engines mounted in the hull driving the pusher propellers via shafts with bevel gears as in previous Oertz designs. Originally, ailerons were mounted only on the forward upper wing. These proved inadequate and an additional pair of ailerons was fitted between the aft pair of wings.

Other than its tandem wings, the W6 closely followed the configuration of earlier Oertz flying boats. The hull was broad-beamed, the engines were hull-mounted driving pusher propellers, and the tail surfaces were carried high above the hull. Originally wing-tip floats were not used, but these were added based on flight experience.

The W6 was significantly larger than earlier Oertz flying boats and carried a crew of three. Only one aircraft, Marine Number 281, was ordered and built.

Oertz W6 Specifications		
Engine:		2x240 hp Maybach Mb.IV
Wing:	Span	20.00 m
	Area	162.7 m²
General:	Length	14.58 m
	Empty Weight	3,780 kg
	Loaded Weight	5,030 kg

Above & Below: Oertz W6 Marine Number 281, in its final configuration after auxiliary ailerons were added between the rear pair of wings and wing-tip floats were installed under the forward lower wing.

Oertz W6 SVK Drawing

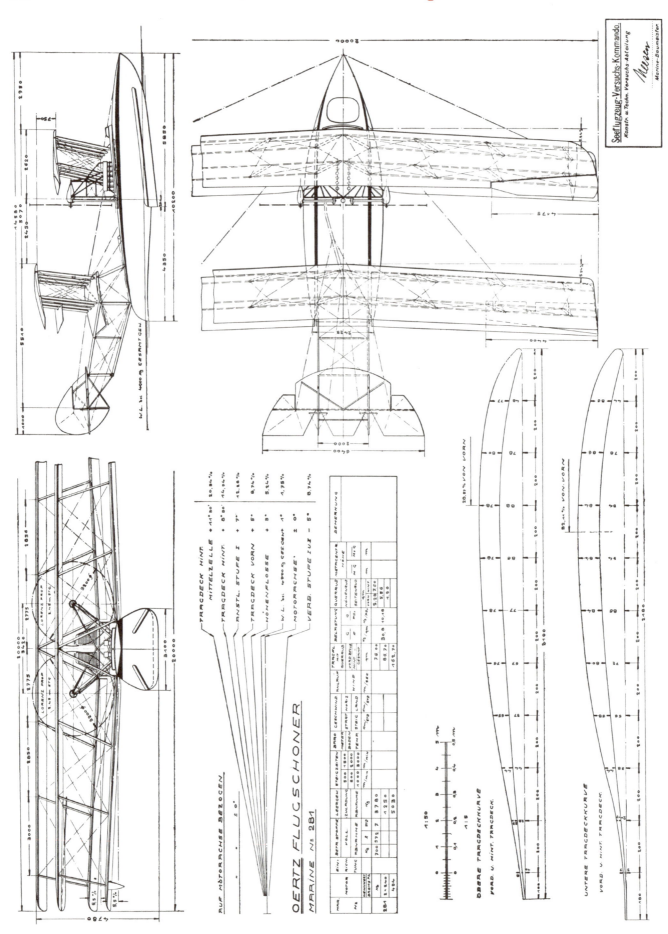

Oertz W6 Hull SVK Drawing

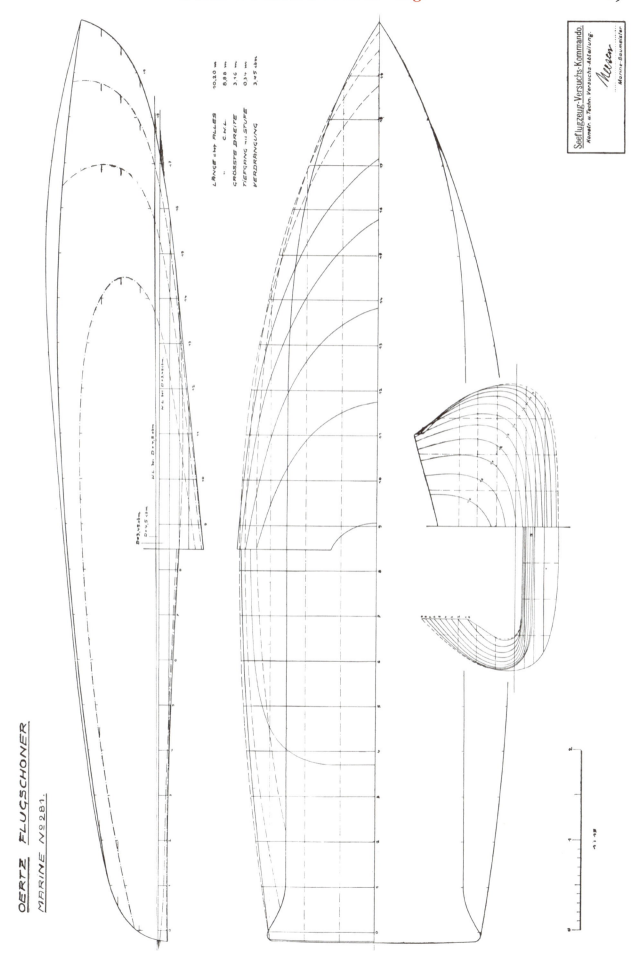

Oertz W7

This Page: Oertz W7 MN 475 *Maruschka* at Borkum in 1916; it was one of two W7 flying boats built. (Photos courtesy Henning Oppermann)

Two examples of the Oertz W7, Marine Numbers 474–475, were ordered in April 1915; they were delivered in May and June 1916. The W7 boats were powered by a 150 hp Maybach Mb.III and were similar in configuration to the W5 but were smaller, two-bay boats.

Oertz W6 Flugschoner

Oertz W8

Above & Below: Oertz W8 Marine Number 1157.

The last Oertz flying boat to be built was the W8. It was known as the *Kruppboot* because it was presented to the Navy by Krupp von Behlen und Halbach. The W8 was powered by a 240 hp Maybach Mb.IV and generally followed the configuration of earlier single-engine Oertz flying boats. However, it reverted to conventional wing-tip floats instead of the sprung boards used in many earlier designs, and additional bracing for the upper wing tips was added. Like the other single-engine Oertz boats, it had a crew of two seated side-by-side. Only one was built.

The photos show the W8 was built with the

Oertz W8 Specifications		
Engine:	240 hp Maybach Mb.IV	
Wing:	Span	19.6 m
	Area	70.2 m²
General:	Length	10.7 m
	Height	3.55 m
	Empty Weight	1,578 kg
	Loaded Weight	2,224 kg
Maximum Speed:		140 km/h

exceptional finish quality typical of Oertz, and its refined design enabled a top speed of 140 km/h, faster than any other Oertz design.

Oertz W8 SVK Drawing

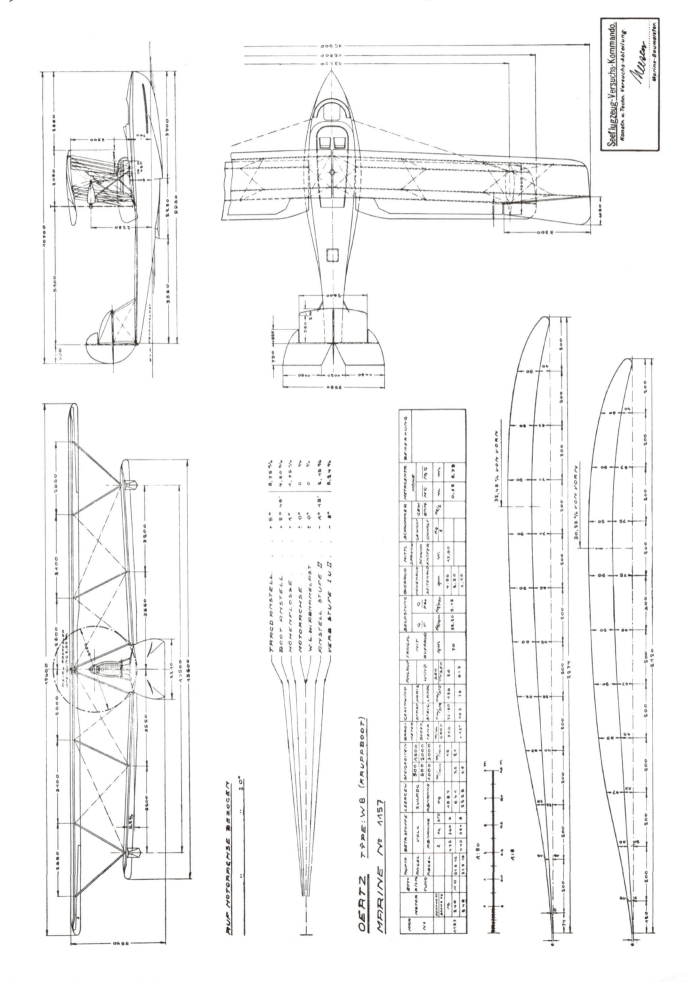

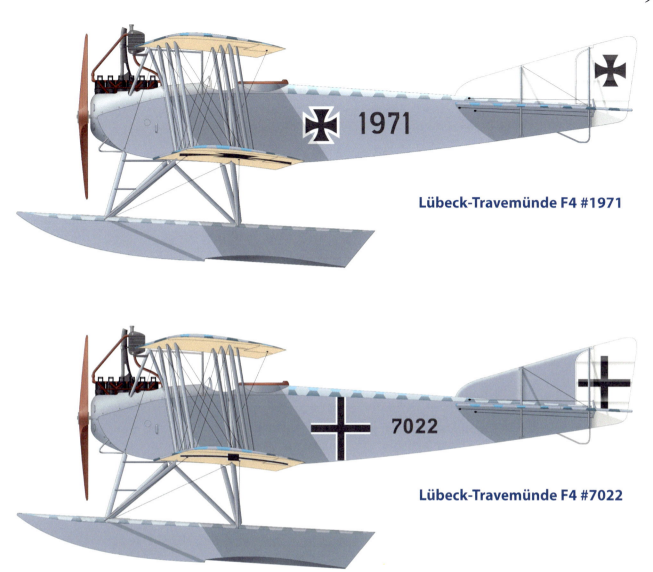

Lübeck-Travemünde F4 #1971

Lübeck-Travemünde F4 #7022

Lübeck-Travemünde F4 #F.46 in postwar Norwegian service

In Retrospect

From the first SF1 design, Sablatnig aircraft generally had streamlined fuselages with good nose entry and small frontal area for minimum drag. Moreover, Sablatnig had the advantages of being a highly-trained mechanical engineer and a successful, well-known pioneer pilot. Therefore it is especially ironic and disappointing that the Sablatnig SF designs were characterized by mediocre performance, flying characteristics, and maneuverability, and by fragile airframes despite a profusion of drag-producing struts and bracing wires. Aircraft design is a particular challenge; the airframe must be strong, yet light and streamlined. Competent engineering can accomplish any two of those qualities, but combining all three requires careful work and ingenuity. Sablatnig designs failed to combine these required qualities; in comparison the Brandenburg W12 and its derivatives succeeded brilliantly.

The Sablatnig designs were not as robust as their Friedrichshafen competitors nor did they possess the performance and combat effectiveness of the Brandenburg two-seat floatplane fighters. Therefore it is surprising that Sablatnigs were produced in quantity given their modest strength and performance. However, it is no surprise that, as a producer of second-string seaplanes, the Sablatnig company disappeared in the harsh postwar economic environment in Germany. Despite their delicate elegance, Sablatnig seaplanes are all but forgotten today, a consequence of their mediocrity.

The story of the handful of floatplanes built by the *Kaiserliche Werften* is even more obscure than that of Sablatnig. Ostensibly the *Kaiserliche Werften* started designing and building training floatplanes to ensure a supply of relatively modern naval training aircraft when the only training aircraft available were retired, worn-out former front-line aircraft. There may have been an element of truth in this, but given the inefficiency of designing and building aircraft in such small numbers, one cannot escape the impression that the main reason the *Kaiserliche Werften* designed and built floatplanes was because the authorities wanted to do so. In particular, if the *Kaiserliche Werften* absolutely had to build modern trainers for the naval air stations, they could have built copies of successful designs by other manufacturers under license. This speculation becomes even stronger with the realization that later in the war some combat types were designed and built, which was certainly not consistent with the official story that the only way to get modern trainers was for the *Kaiserliche Werften* to build them. Overall a total of 22 aircraft of seven different designs were ordered and apparently built. This must have been an interesting hobby for a group of aviation enthusiasts at the *Kaiserliche Werften* but had no impact on the naval air war.

A subsidiary of DFW, Lübeck-Travemünde produced a small number of robust reconnaissance floatplanes that served well in the northern waters around Germany from the second half of 1917. The designs were conventional for the time and supplemented Friedrichshafen's better-known designs near war's end. The Lübeck-Travemünde F4 continued to give good postwar service in non-military roles.

Although Luft Torpedo Gesellschaft (LTG) was established to develop air-launched torpedoes, for some reason a single-seat floatplane fighter was ordered in prototype quantities despite the fact that adequate designs from more established manufacturers were already in service. The low-powered FD 1 prototypes had no chance of being ordered into production and nothing came of this effort.

Max Oertz was famous for his yacht designs before the war and started building flying boats in 1913. The Oertz flying boats were cleverly designed, very well built, and offered excellent seakeeping. However, production deliveries were painfully slow and the German Navy preferred floatplanes to flying boats, so few Oertz flying boats were ordered, and these interesting waterbirds had very little impact operationally.

Seaplanes, especially floatplanes, were built in small production batches by a number of German aircraft manufacturers. The five little-known manufacturers covered in this book produced a variety of designs, only a few of which were placed in production. Of these, only the floatplanes from Sablatnig and Lübeck-Travemünde made any significant impact on the naval air war. The German Navy relied most on the proven, sturdy reconnaissance floatplanes from Friedrichshafen and the fast, maneuverable two-seat floatplane fighters from Brandenburg. Accordingly, it is the Friedrichshafen and Brandenburg floatplanes that are best remembered today for fulfilling the bulk of the German Navy's air combat requirements.

German Seaplanes of WWI

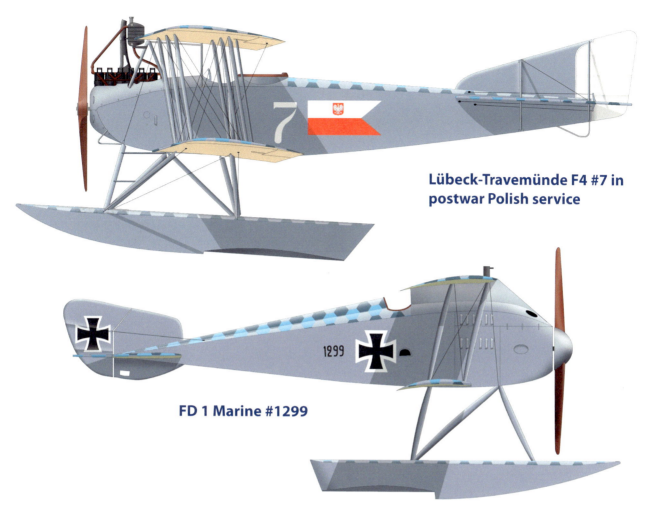

Lübeck-Travemünde F4 #7 in postwar Polish service

FD 1 Marine #1299

Bibliography
Books
Gray, Peter, and Thetford, Owen, *German Aircraft of the First World War*, second revised edition, New York: Doubleday & Company, Inc., 1970.

Nowarra, Heinz J., *Marine Aircraft of the 1914–1918 War*. Letchworth, Harts: Harleyford Publications, 1966.

Articles
Leaman, Paul, "The Oertz Flying Boats" *Cross & Cockade International* Vol.42 No.3, Autumn 2011, p.200–207.

"The Oertz Flying Boats, *Flight*, Oct. 9, 1919, p.1345–1349.

Internet
Sablatnig Flugzeugbau GmbH, www.aviastar.org/manufacturers/1827.html
Sablatnig, Joseph, de.wikipedia.org/wiki/Joseph_Sablatnig
http://de.wikipedia.org/wiki/Max_Oertz
en.wikipedia.org/wiki/Kaiserliche_Werften
en.wikipedia.org/wiki/Kaiserliche_Werft_Danzig
en.wikipedia.org/wiki/Kaiserliche_Werft_Danzig_404
en.wikipedia.org/wiki/Kaiserliche_Werft_Danzig_467
en.wikipedia.org/wiki/Kaiserliche_Werft_Danzig_1105
en.wikipedia.org/wiki/Kaiserliche_Werft_Danzig_1650
de.wikipedia.org/wiki/Kaiserliche_Werft_Kiel
en.wikipedia.org/wiki/Kaiserliche_Werft_Kiel_463
en.wikipedia.org/wiki/Kaiserliche_Werft_Wilhelmshaven
en.wikipedia.org/wiki/Kaiserliche_Werft_Wilhelmshaven_401
en.wikipedia.org/wiki/Kaiserliche_Werft_Wilhelmshaven_461
en.wikipedia.org/wiki/Kaiserliche_Werft_Wilhelmshaven_945
en.wikipedia.org/wiki/Kaiserliche_Werft_Wilhelmshaven_947

Index

Name	Pages
Harder, Hans	74
Heinrich, Prinz	17
Oertz, Max	74, 94
Plüschow, Wolfgang	17
Sablatnig, Joseph	3, 7, 9, 21, 94
Viktoria, Augusta, Kaiserin	17

Sablatnig SF2

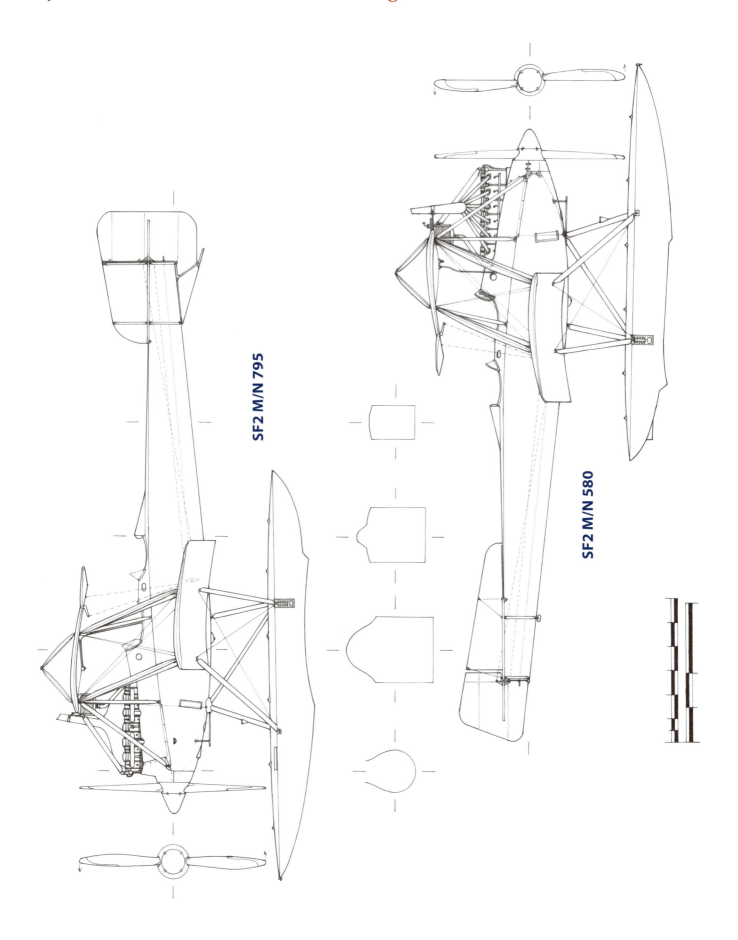

SF2 M/N 795

SF2 M/N 580

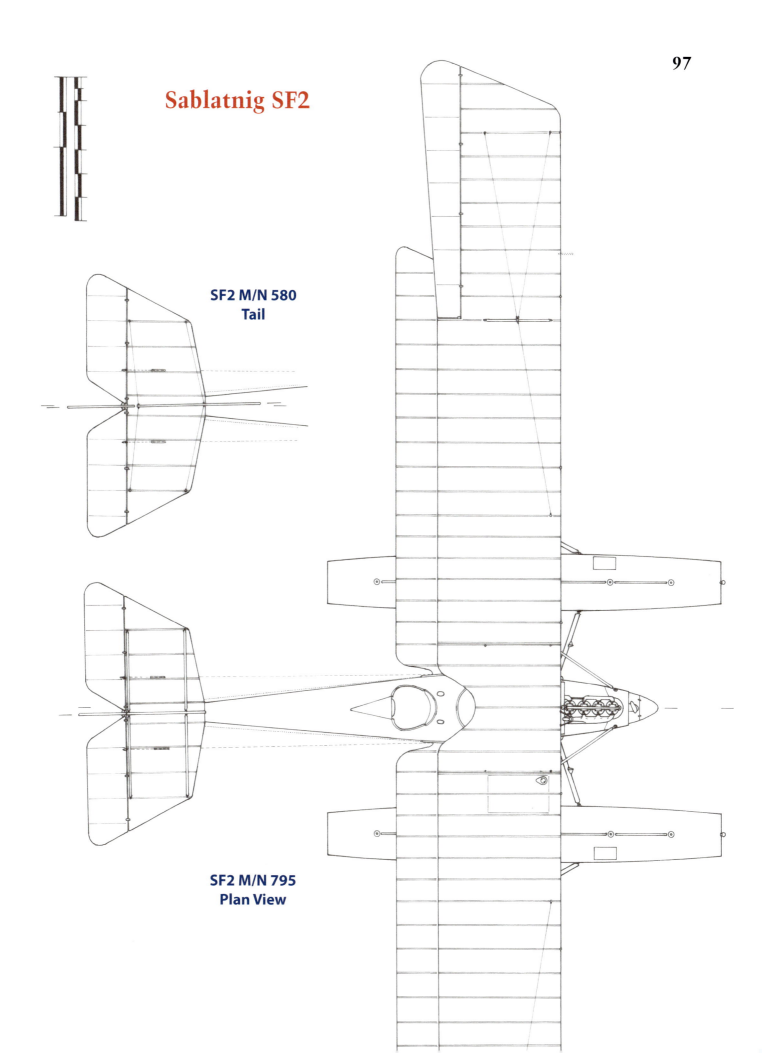

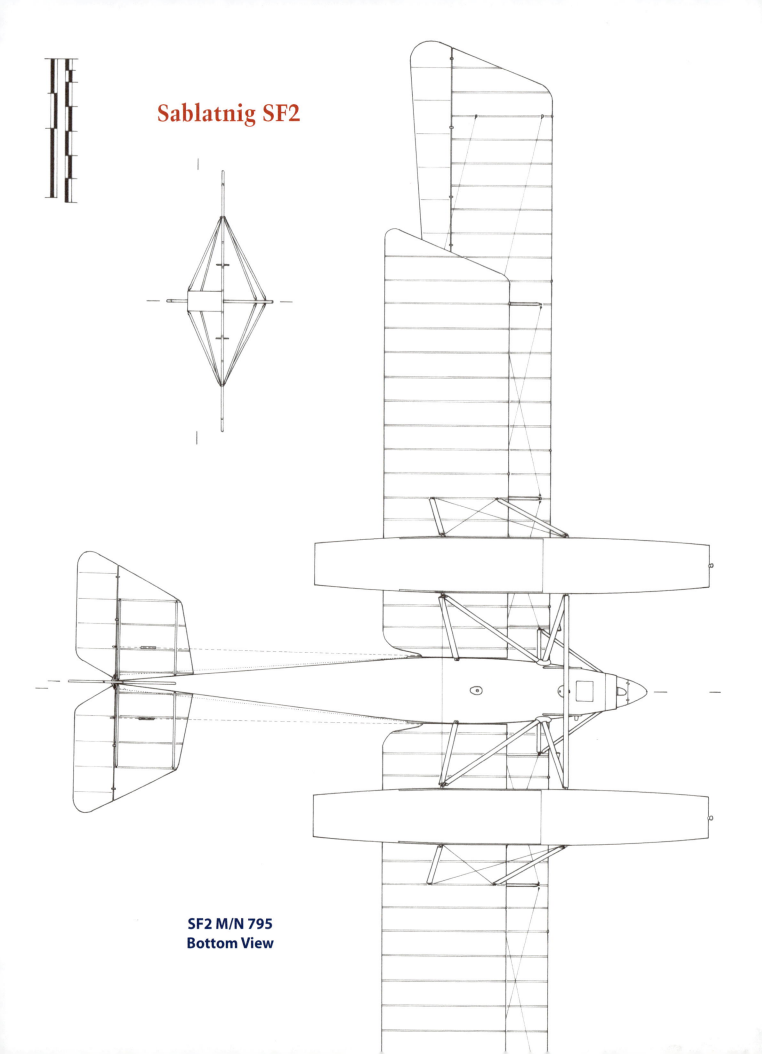

Sablatnig SF2

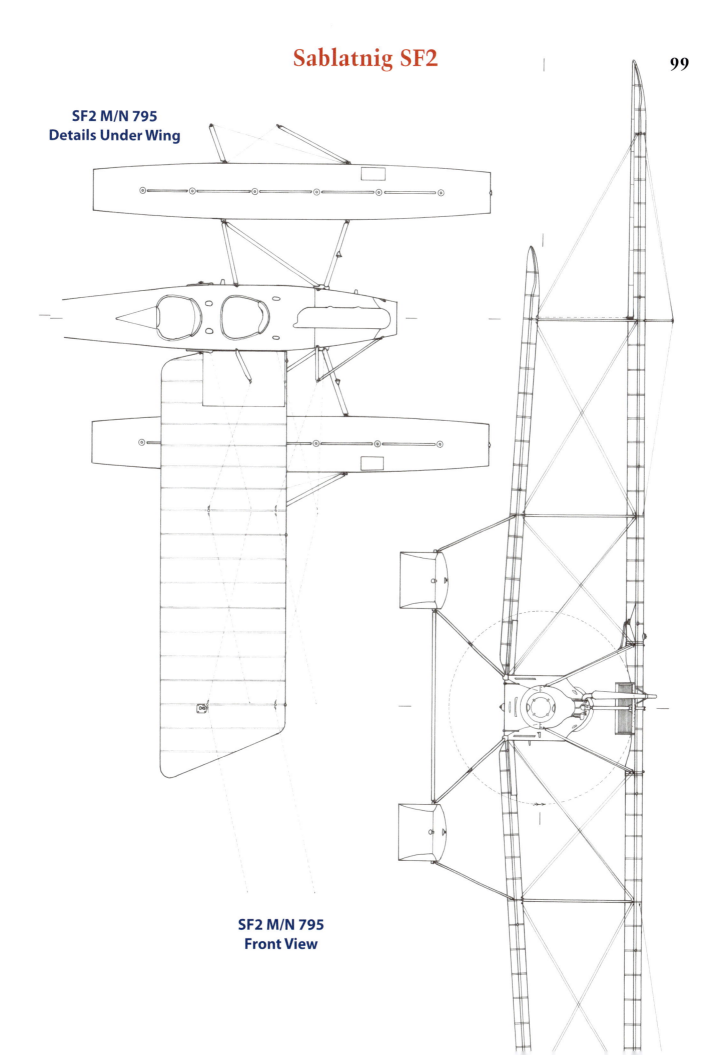

SF2 M/N 795 Details Under Wing

SF2 M/N 795 Front View

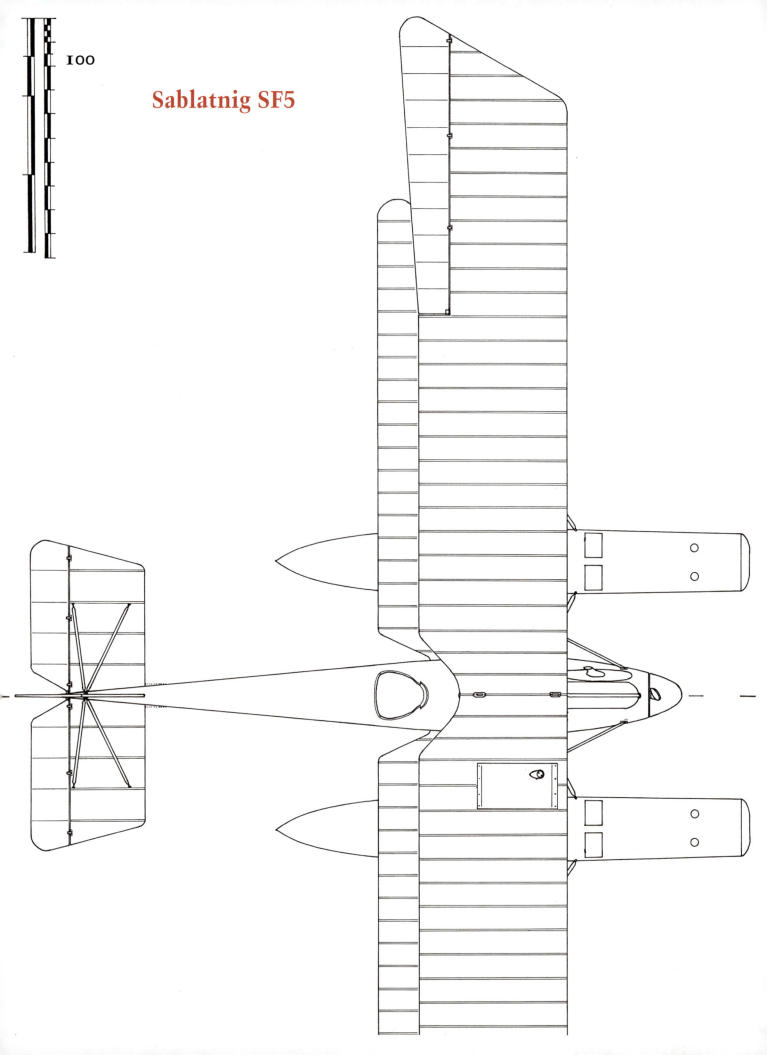

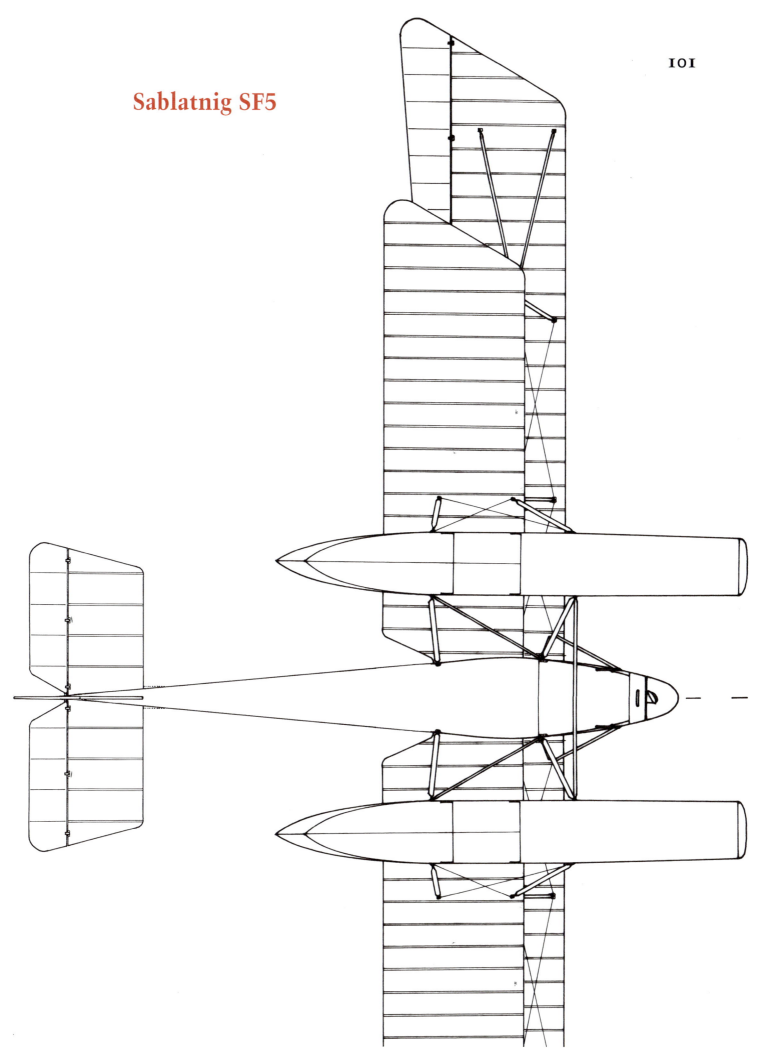
Sablatnig SF5

Sablatnig SF5

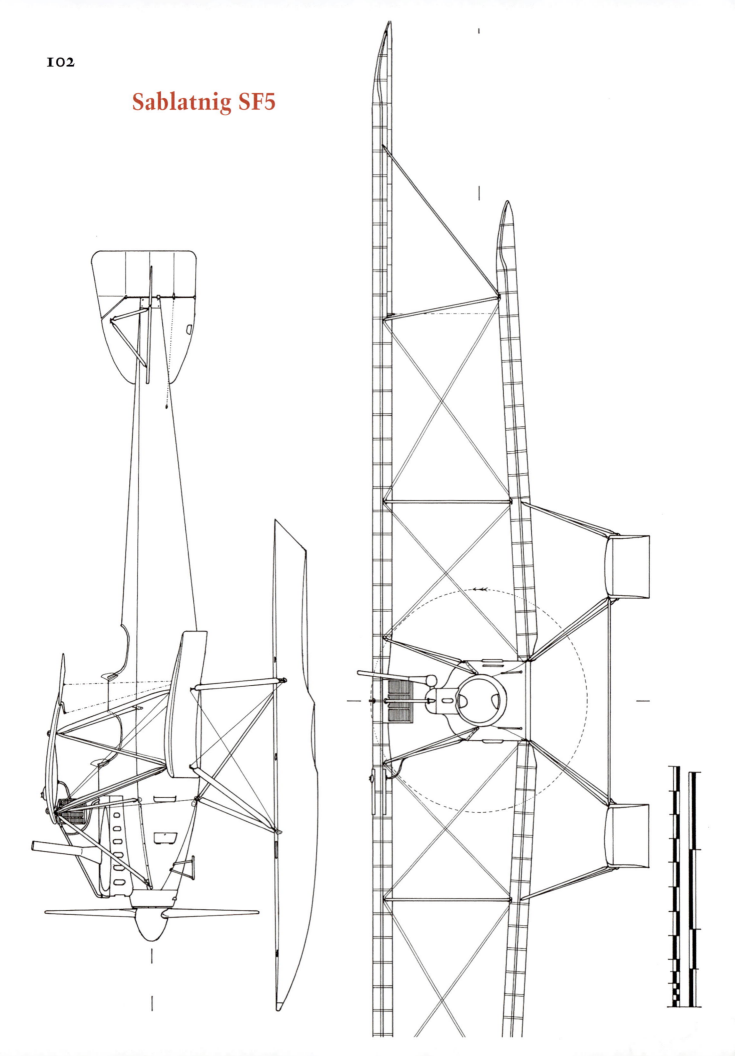

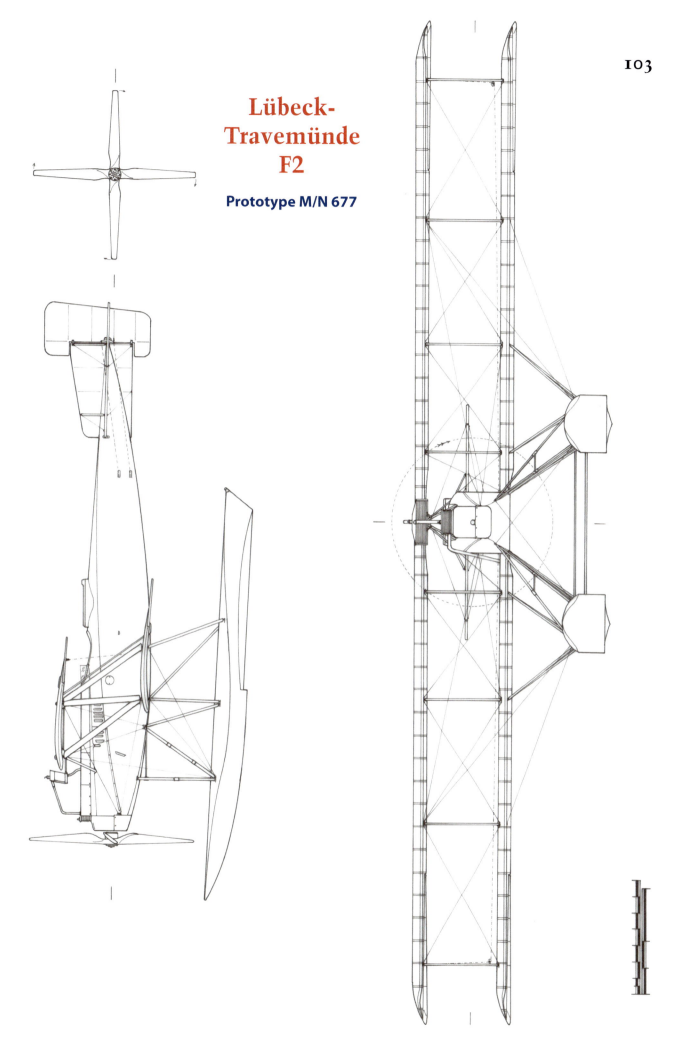

Lübeck-Travemünde F2

Prototype M/N 677

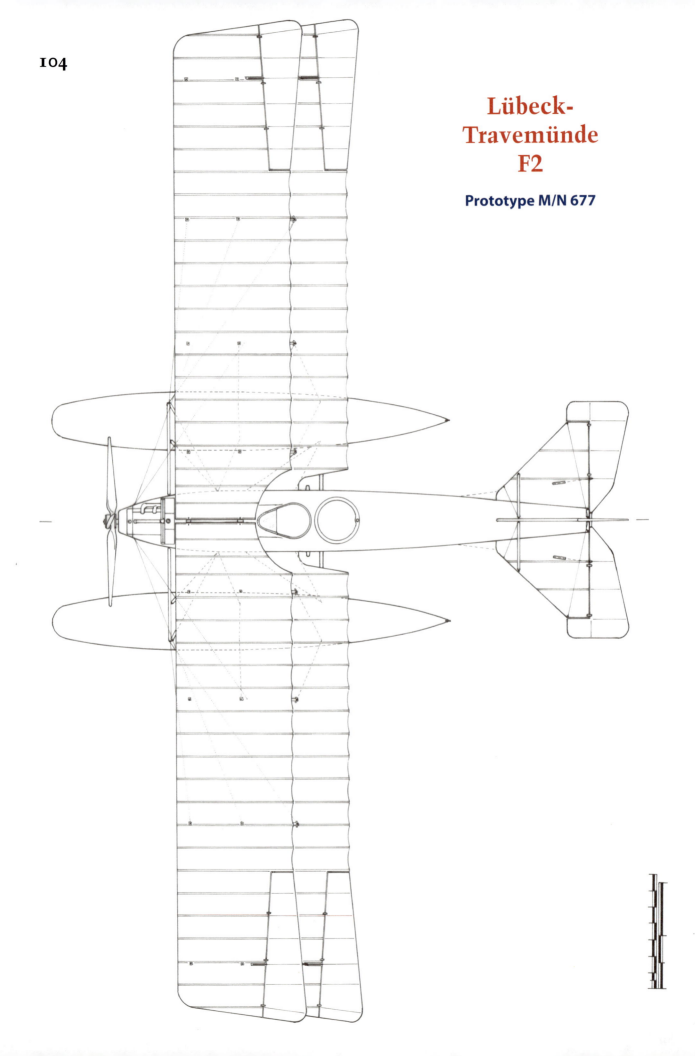

Lübeck-Travemünde F2

Prototype M/N 677

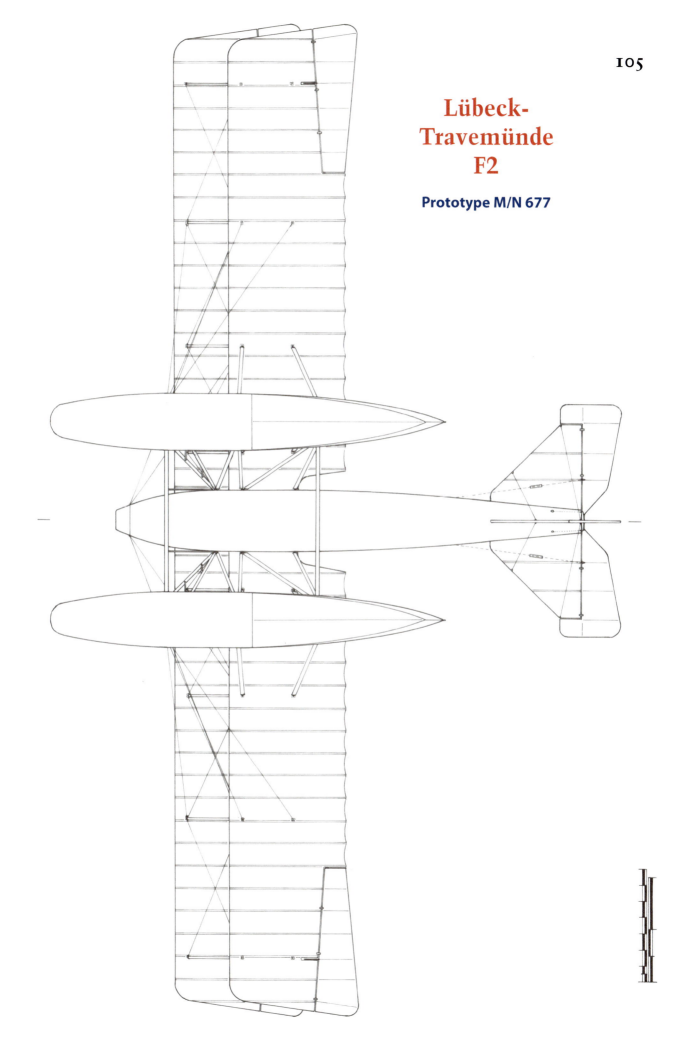

Lübeck-Travemünde F2

Prototype M/N 677

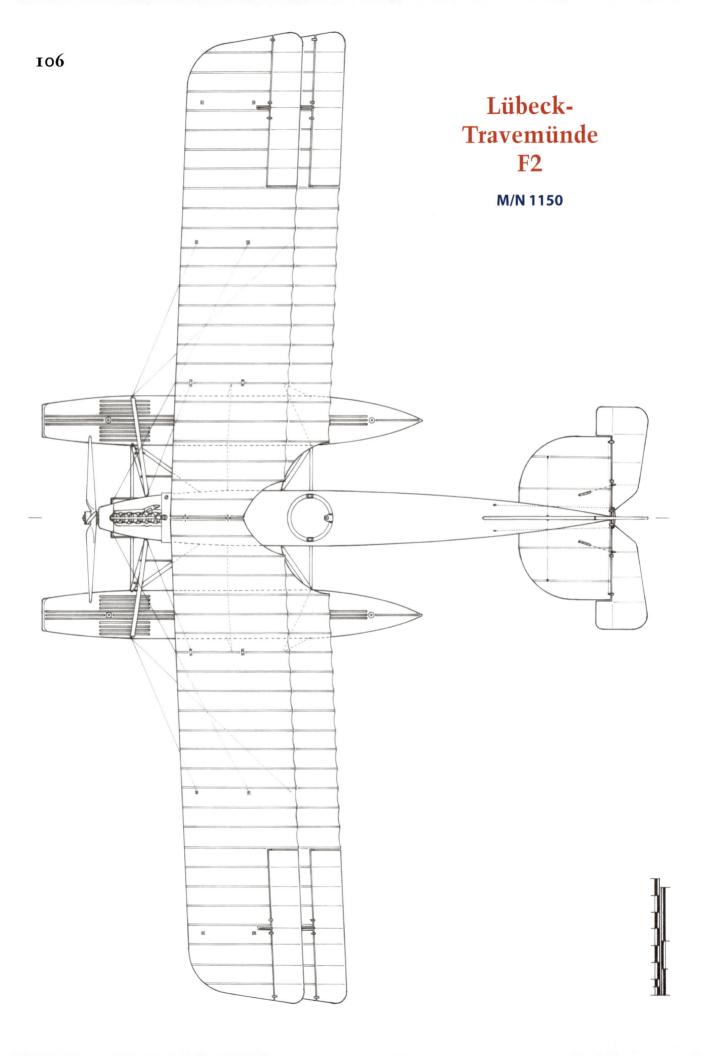

Lübeck-Travemünde F2

M/N 1150

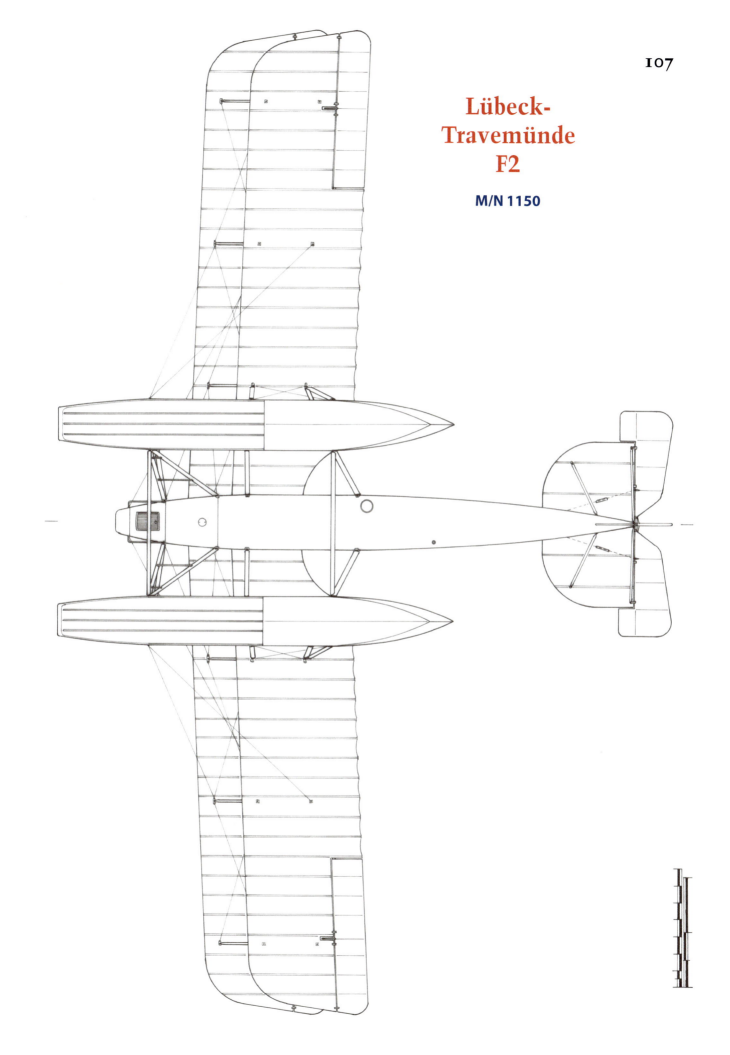

Lübeck-Travemünde F2

M/N 1150

Lübeck-Travemünde F2

M/N 1150

Lübeck-Travemünde F4

M/N 1971

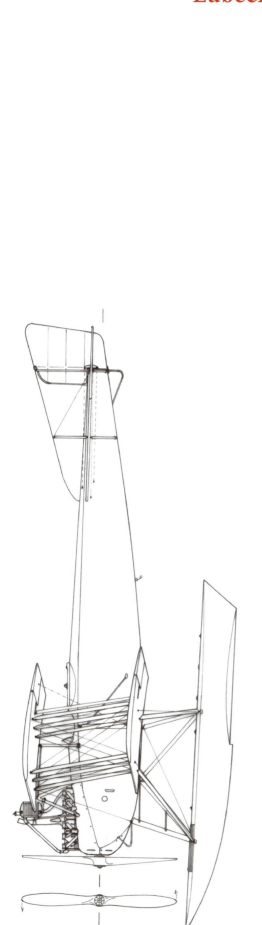
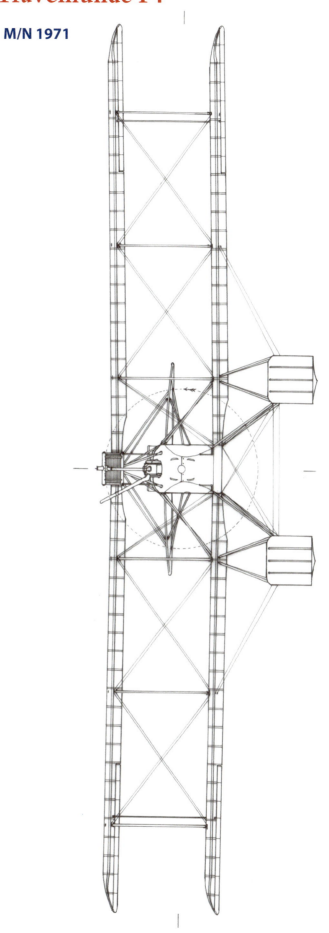

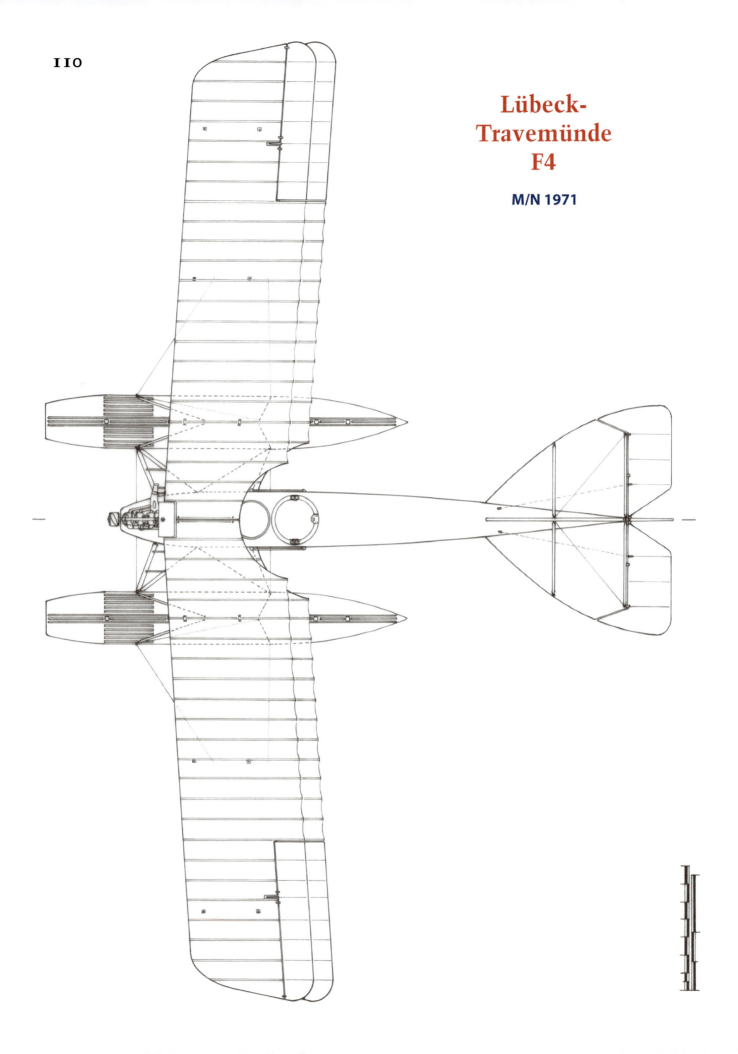

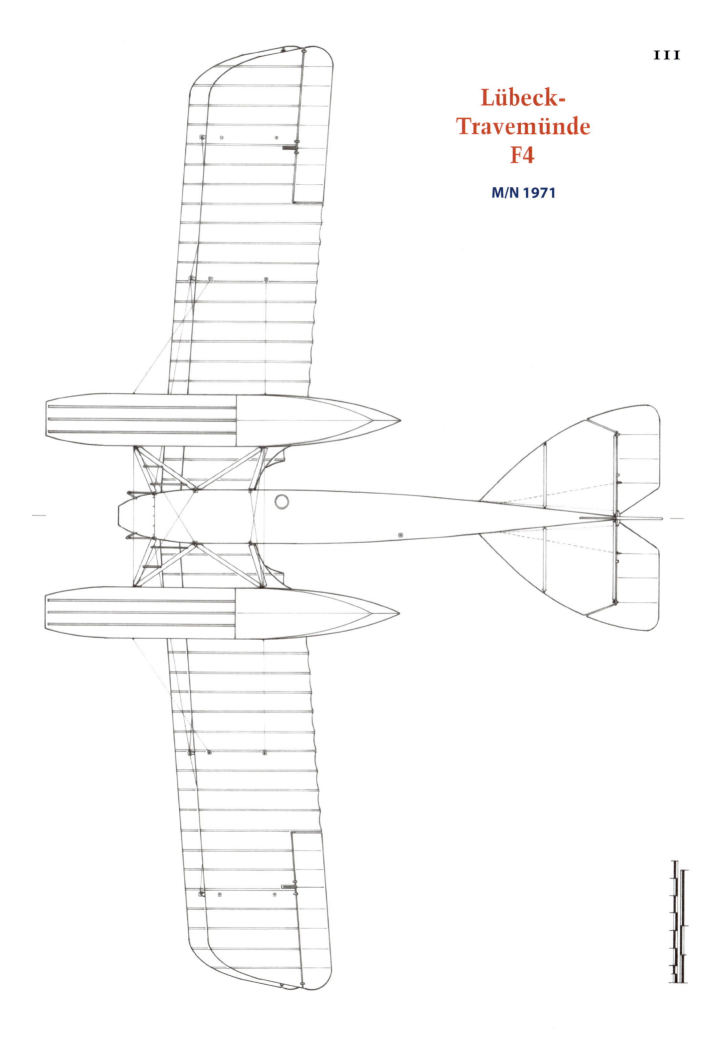

Lübeck-Travemünde F4

M/N 1971

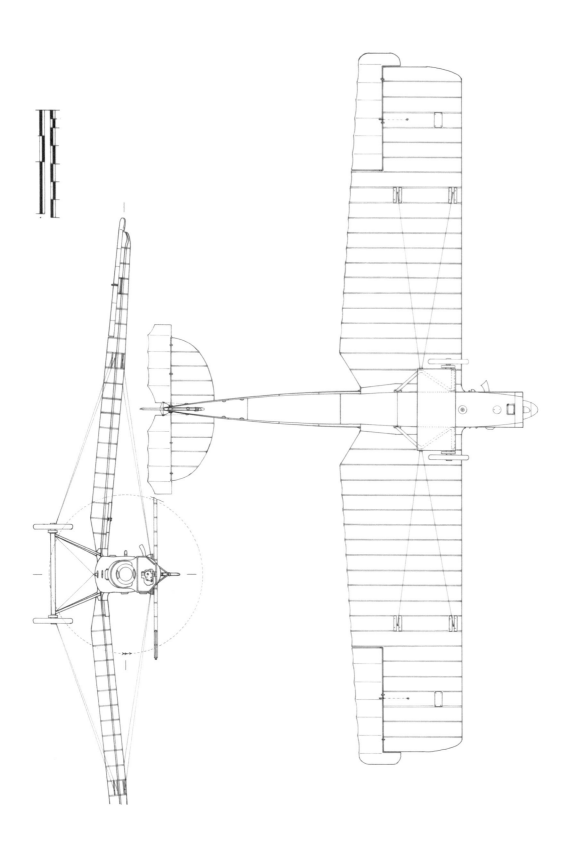

Sablatnig C.III

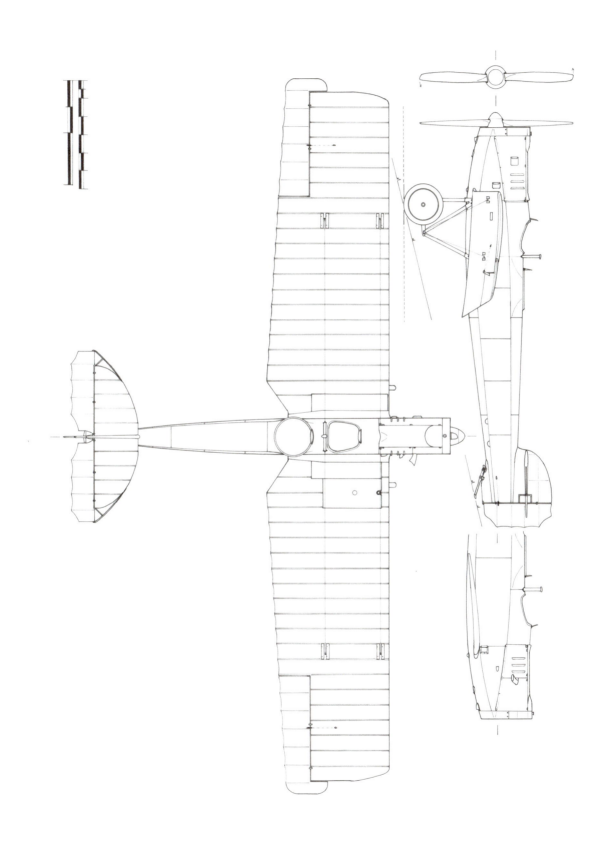

Afterword

Above: Postwar this Lübeck-Travemünde F4 was converted to passenger service; a passenger cabin was added in the rear fuselage as shown. It was a Norwegian aircraft owned by the United Sardine Factory.

The two F4 floatplanes that served in Norway were apparently both imported by the United Sardine Factory, Bergen. The first was imported and registered in August 1919 and may have been ex German D-73. The serial number was given as 503. This machine received the registration N-2. Returned to Germany in the spring of 1920 for modification, it went to the Norwegian Navy as F.46. A second F4, civil registration N-15, was also imported by the United Sardine Factory in July 1920, but was sold to the Navy as F.48, finally being scrapped on 7 April 1930.

Johan Høver recalled that the Norwegian Navy purchased two F4 floatplanes. Friedrich Christiansen, the German naval ace, arrived at Horten in July 1920 in order to clear the way through customs for these two F4 floatplanes. The aircraft had been converted to take four passengers and were for a proposed sale in Bergen. The sale did not go ahead and their return through customs was postponed. In the event the Norwegian Naval Air Service was able to buy the two aircraft for a reasonable price party due to the favorable exchange rate for the German Mark then prevailing. These aircraft were equipped with a 220–260 hp Benz six cylinder water-cooled engine. This engine had a good reputation for reliability and could be bought cheaply from Sweden. Given the Navy serials F.26 and F.48, they were later converted to three seaters with a spacious cabin for the radio operator The design was so efficient in its lifting capacity that in 1923, F.46 was converted to take a torpedo. It could easily take off with a 500 kg V-b torpedo and completed a lengthy torpedo program. As a result thereof the Naval Air Service and Mine Section gained a lot of useful experience in this field. During practice runs 75% hits were achieved. The F4 type served until 1930.

In 1920 the Norwegian airline Der Norsk Luftfartrederi AS (DNL) was operating an aerial service to Bergen, Haugesund, and Stavanger on the west coast of Norway. The loss of Supermarine Channel I flying boat N-11 forced them to purchase the ex-Navy Lübeck-Travemünde F4 (serial F.46). This service was closed down soon afterwards on 15 October 1920, and the F4 was returned to the Navy.

Notes on the Norwegian Naval Flying Boat Factory by Erik Hiedesheim via Colin Owers.

Made in the USA
Columbia, SC
02 January 2025

51092572R00064